Water Chemistry

Water Chemistry

Oliver Jenkins

Larsen & Keller
www.larsen-keller.com

Water Chemistry
Oliver Jenkins
ISBN: 978-1-64172-116-5 (Hardback)

© 2019 Larsen & Keller

▤ Larsen & Keller

Published by Larsen and Keller Education,
5 Penn Plaza,
19th Floor,
New York, NY 10001, USA

Cataloging-in-Publication Data

Water chemistry / Oliver Jenkins.
        p. cm.
Includes bibliographical references and index.
ISBN 978-1-64172-116-5
1. Water chemistry. 2. Geochemistry. 3. Hydrology. I. Jenkins, Oliver.
GB855 .W38 2019
551.48--dc23

For more information regarding Larsen and Keller Education and its products, please visit the publisher's website www.larsen-keller.com

Table of Contents

Preface

The identification and quantification of the properties and components of water samples are under the domain of water chemistry. The type and sensitivity of tests performed depends on desirability and intended use of water. It integrates studies in water quality, hydrology, pollution and geothermal waters. Methods of mass spectrometry and gas chromatography, specialized organoleptic methods, etc. are routinely used in water chemistry. Such studies are crucial in the domains of forensic analysis, research, drinking water supply, etc. This textbook is a valuable compilation of topics, ranging from the basic to the most complex theories and principles in the field of water chemistry. Most of the topics introduced herein cover new techniques and applications of this field. For someone with an interest and eye for detail, this book covers the most significant topics in this area of study.

A short introduction to every chapter is written below to provide an overview of the content of the book:

Chapter 1, The hydrosphere is the sphere of the Earth, which comprises of the water on, above and below the surface. The topics elucidated in this chapter cover some of the important aspects of the hydrosphere and Earth's water reservoirs, such as fresh water, ground water and polar ice caps; **Chapter 2**, Water is a polar inorganic compound, which is tasteless and odorless at room temperature. It is the most abundant substance found on Earth in the form of solid, liquid and gas. The principal properties of water like taste and odor, color and appearance, polarity and bonding, etc. have been carefully analyzed in this chapter; **Chapter 3**, Water is represented by the chemical formula H_2O, which means that each water molecule consists of one oxygen atom that is bonded to two hydrogen bonds through covalent bonds. This chapter closely examines the chemical composition of water. It includes topics like chemical bonding found in water, Vienna Standard mean ocean water, heavy water, light water and doubly labeled water, among others; **Chapter 4**, The properties and chemical components of water are studied under water chemistry. An extensive study of the fundamental concepts and principles of water chemistry, such as metal aquo complexes, hydrolysis, water of crystallization, hydration reaction, electrolysis, dehydration reaction, etc. is vital to the understanding of water chemistry which have been covered in elaborate detail in this chapter; **Chapter 5**, The mass of water remains more or less constant over time on Earth. Water moves from one reservoir to the other due to physical processes of evaporation, precipitation, condensation, surface runoff, etc. This is an important chapter, which will analyze in detail about the water cycle and the physical processes influencing the water cycle.; **Chapter 6**, The contamination of water bodies, such as lakes, rivers, aquifers, oceans and groundwater, due to anthropogenic activities is termed as water pollution. The topics included in this chapter such as effluent, acidification of water, ocean deoxygenation, etc. addresses the important facets of water pollution.

Finally, I would like to thank my fellow scholars who gave constructive feedback and my family members who supported me at every step.

Oliver Jenkins

The Hydrosphere

The hydrosphere is the sphere of the Earth, which comprises of the water on, above and below the surface. The topics elucidated in this chapter cover some of the important aspects of the hydrosphere and Earth's water reservoirs, such as fresh water, ground water and polar ice caps.

Water

Water is largely discussed from a physical or a normative perspective. From a physical perspective, discussions are usually limited only to the terrestrial part of the hydraulic cycle, comprised of blue water, green water and gray water. Much of the discussion focuses on blue water, differentiating between surface water and groundwater. Most discussions of water thus focus on less than 3% of the total water of the hydrosphere. Yet, while these discussions are couched in technical terms, they reflect deeper societal issues, pertaining to values and norms. Hence, it should not be surprising that in these discussions different actors espouse different views of water. Many view water as a natural resource. Others (mostly economists) argue that it should be viewed as a commodity or a factor of production. However, water is first and foremost a source of life. Thus it has been argued that water should be considered as a basic need, and therefore it constitutes a right to which people are entitled. Proponents of bio-centric approaches have argued that similar entitlements should also be extended to other species.

These conflicting normative views of water have practical ramifications, as each is associated with a different management approach. Thus, the often acrid arguments between these views are closely intertwined with conflicts over concrete policy measures. A common feature in these normative discussions is their view of water as a unitary substance. That is, participants refer to all the (usually blue) water as 'water'. Yet, this has not always been the case. The view that water is a single substance, due to its chemical composition, is modern. In the past, water was viewed in the plural term, 'waters', constituting a variety of wet cool substances such as light water, standing water, clear water etc.. While the fact that water has a common chemical composition is not contested, it is possible to ask whether all the liquids composed of H_2O should indeed be treated equally. If the answer is negative, as I propose here, then it is possible perhaps to resolve some of the differences in normative views of water by suggesting that attitudes towards water need not subscribe to one normative view, as different views may pertain to different waters.

If water can be differentiated, two immediate questions arise: According to which criteria should water be differentiated? How can this differentiation affect policies? To this end, first discuss the use of water, beginning with the question of what are the basic needs and hence rights of humans. Then look at the needs of nature, before coming back to humans. Then, discuss the role of water in human spirituality and as a factor in maintaining communities as such. In the fourth and fifth sections, look at the questions of water for the production of food and at other water uses, arguing that these uses should be viewed as economic demands, rather than normative needs. After discussing the policy and pricing implications of the differentiation between waters on the basis of use, go on to ask whether waters should also be differentiated according to source. Answering in the affirmative, suggest a normative differentiation of water based on both use and source.

The Human Need for Water

Water is necessary for human survival. This well-recognized fact is the basis for the emerging view of basic human water needs as a human right to water. But human rights are about more than survival. They include dignity for, as Hiskes states: 'Human dignity is both producer and product of moral rights, in the sense that morality is impossible without them'. Human dignity also underlies the Millennium Development Goals' premise that extreme poverty is unacceptable because it does not allow for dignified living. But the water-related goals stated in the MDG pertain only to access to clean water and sanitation, without specifying what is meant by access, or the amounts to which humans are entitled. The unanswered question is thus: How much water constitutes the amount that is sufficient to assure human dignity?

There is a wide variance of estimates of basic needs, not all of them relating to the concept of dignity. Thus, while Peter Glieck argues that 50 liters per capita per day constitute the basic need for humans, Falkenmark, for example, suggests that 274 L/c/d are needed for a good standard of living.

Chenoweth, who reviewed the literature on basic water needs, identified two approaches for estimating minimum water needs. The first is based on specific quantities required for basic domestic functions. This is the approach used by Glieck, as well as many others. The second is based on observations of the amount of water used by prudent societies that enjoy a high level of development. This is the approach used by Falkenmark and Shuval. While the first approach tries to identify the basic minimal water needs, but does not tackle the question of dignity analytically, the second approach begins by identifying societies that do offer a dignified level of living, searching for the minimum threshold of water usage that is required to sustain a high level of human development. Chenoweth termed this second approach the development-efficiency approach, whilst he termed the first a sectoral approach. Both approaches implicitly assume all people behave similarly. However, societal and climatic differences may lead to variances in the quantity of water used for various activities (for example the amount of water used for food preparation may be affected by the extent to which people eat at home or out of home), and different countries have different non-agricultural economies, and hence water use patterns.

Based on these approaches, Chenoweth finds that a country can meet its domestic water requirements and maintain a non-agricultural economy that is capable of sustaining a high level of development with as little as 120 L/c/d. Adding low (but not excessively so) losses in the water system to this, he came to the conclusion that 135 L/c/d, which is akin to 50 m³/c/yr, can be viewed as the minimal threshold required for social and economic development that permits the achievement of high human development.

Taking Chenoweth's figures as a base, and following a workshop in which leading experts discussed the issue, Feitelson et al. suggested that in the Middle East, due to the greater aridity and perhaps slightly higher water losses, a figure of 60 m³/c/yr should be considered as a normative domestic water need. Taking into account the variations in domestic use noted above, it seems that the quantity of water that can be considered as a normative need, and hence a basis for establishing a human right for water that allows for a dignified level of living, can be capped at 60 m³/c/yr. This quantity is significantly lower than the quantities Falkenmark and Shuval quoted on the basis of the Israeli experience, but also significantly higher than the basic needs defined by Glieck.

Water for Nature

The relationship between humanity and natural systems is discussed in the Bible, in the book of Genesis. However, different views are expressed within Genesis. In the first chapter, Man is viewed as supreme, and thereby allowed to subdue the earth, and to dominate the fauna and flora for his benefit. In the second chapter, a different normative approach is advanced, whereby Man is a steward of nature, who should take care of it. This second normative position is the basis for the evolving ethic whereby humans have a moral obligation to sustain natural ecosystems and hence to retain water in streams, rivers, aquifers and lakes for this purpose. This ethic may be couched in the language of stewardship, such as Postel's water ethic, or Falkenmark & Folke's ecohydrosolidarity; in a bio-centric ethic, whereby humans are only the first among equals; or in a view of sustainability and inter-generational ethics, whereby the environmental rights are part of a human right for a green future.

The quantities of water that should, normatively, be allocated for nature are more difficult to define than the human needs discussed above. Clearly, any use of water by humans that is based on the abstraction of water from natural courses or aquifers affects the environment. Hence, the amount of water that should be retained in nature is not the same as the amount that was there in a pristine state. Rather, the focal point should be on the quantities and chemical composition of water that is necessary for the natural systems to survive. From an ethical perspective these quantities can be seen as similar to the basic needs of humans discussed above. In rivers, this minimal threshold may be referred to as the minimal environmental flow that should be retained in the river, or returned to it.

However, the quantity and quality of water needed to maintain ecosystems are highly variable across space and time, as a function of environmental, ecological and climatic variables, as well as of human perceptions and attitudes. Moreover, ecosystems are dynamic and feedback mechanisms produce non-linear results. Thus, while clear thresholds exist, it is very difficult to identify them in advance. Moreover, due to the spatiality of water–ecology–environment dynamics, it is impossible to come up with definite globally-applicable thresholds as was attempted with regard to human needs. Rather, all that can be said is that there is water that should be viewed, normatively, as water that is necessary for the maintenance of healthy and viable ecosystems. But the quantity and quality of this water has to be determined on a case by case basis, taking into account both local and regional perturbations. It should be noted that in the case of nature, it is insufficient to define the water needed to maintain a healthy and viable ecosystem in terms of quantity and chemical composition, as the rates of water flow may be no less important. Thus, in the case of water for the environment, the pattern of water flows should be included as an additional dimension. Still, the minimal flows do not supply the full array of environmental services that aquatic ecosystems may supply. But additional water, beyond the minimal requirements, will have to be balanced against other demands. To this end the various valuation techniques developed to value the environmental services of water in monetary terms may be used.

The Unique Needs of Humans

In addition to their greater capability to manipulate water, humans differ from all other species in the variety of their needs. Beyond physical needs, discussed above, humans have spiritual and social needs for water. These should also be accounted for in normative discussions of what constitutes water.

The Spiritual needs of Humans

Water has symbolic properties in many cultures. In many religious tracts, water is viewed as cleansing not only the body but also the soul, as well as being seen as a symbol of purity. As a result, water is used in a variety of religious and cultural ceremonies and traditions for ritual washing or immersion, as well as for cleansing the dead or certain parts of the body. In some cases, specific water bodies, such as the Jordan River, the Ganges and various springs, are viewed as holy or as deities.

The quantity of water used in religious ceremonies or traditions is usually miniscule. However, there may be other attributes of water which are important, such as cleanliness, clarity, or its natural flow. In the case of water bodies imbued with holiness, the continuous flow of water in them has a spiritual meaning for believers. From a normative perspective, this water use is fundamentally different from all other water uses, as it is not related to the physical well-being of people. Hence, it should be regarded as a different kind of water, regardless of the quantities used or the exact rate of flow in holy water bodies.

Human Social needs

Humans are social creatures. We live in communities. Some thinkers have elevated the community to the level of an ideal. The community scale has also been advanced as a normatively-desirable scale of water management. Thus, if water is a requisite for maintaining a community, it is, arguably, a different human need than all those noted above.

The question, in this case, is under which circumstances is water necessary to maintain a viable community? In much of the arid and semi-arid developing world, access to secure water for irrigation is seen as a key tool for addressing the social roots of instability. Hence, it can be assumed that such water should be found only in communities based on subsistence farming.

But the idea of a community as a central building block of society is not limited to societies whose economies are based on subsistence agriculture. Actually, such ideals are derived from Kropotkin's image of advanced Swiss villages. Thus, it should be asked whether water may also constitute a basis for communities in more advanced economies.

Feitelson in an attempt to define water needs for the Israeli–Palestinian case, suggest that water should be considered as a need in the case of 'peripheral' agriculture. Peripheral agriculture is defined by them as agriculture on which communities that are beyond commuting distance from major cities and towns are based. These are communities that have no alternative non-agricultural employment opportunities. Under these circumstances, water for farming is essential for the continued viability of the communities, even if the product is geared toward the market and is not limited to subsistence.

The principle of peripheral agriculture can clearly be extended to circumstances beyond the Israeli–Palestinian case. However, the specification of the quantity of water that should be considered as a social need in this case is likely to be contested. Essentially, the quantity that can be considered a need is the minimal quantity needed to keep farming viable in these communities. Any additional water that is used for farming in these communities should not be seen as any different from the water used to produce food elsewhere.

From a policy perspective, community needs for water are likely to be relevant mainly in arid and semi-arid environments, where farming is dependent on irrigation. In other environments where farming is largely based on green water (that is, farming is largely rain-fed), community needs for water, while existing, may be largely beyond the policy arena where water allocations are determined.

Water for Food Production

Water is often seen as essential for food security. But food is increasingly supplied through the global market. As a result, the main flows of water can be argued to be the flows of virtual water embedded in foodstuffs. These flows of virtual water balance out the discrepancies between the concentrations of humanity and the distribution of land and of the water resources needed for food production. Thus, food security is meaningful today mostly at the global level, and not on a national or local level.

The implication of the reliance on global markets for food security is that the water used to produce food serves essentially as a factor of production, except for the cases noted above where communities rely on subsistence farming or have no alternative employment base. If the water used for food production is viewed as a factor of production it should be discussed in the realm of the market. In this sense it is no different than water used as a factor of production in other economic activities, such as tourism, power production or industry. Some qualifications may apply to this general assertion in cases where the purchasing power of a country is insufficient. However, the adaptations to such shortfalls go well beyond the questions of water allocation, as the most effective of these may be changes in diets toward less animal-based products.

Other uses of Water

The demand for water by humans is greater than the total human needs defined above (including spiritual and social needs). This additional demand is normatively different from the needs identified above. While the needs can be argued to be necessary for the dignified living of people, for maintaining viable ecosystems, for addressing peoples' spiritual needs and maintaining human communities as such, these additional demands constitute desires that people would like to satisfy. Such desires are certainly legitimate. But in essence they are no different from the desires to derive utility from various other goods and services. Thus the demand for these water services has to be balanced against the other desires of the households or municipalities, under their pecuniary budget constraints. In other words, the additional demand has to be treated as any other economic demand, and thus the water that is used to supply this additional demand should be viewed as a commodity.

The abstraction of water entails both direct costs and negative externalities. Thus, the prices that consumers will pay for the water services that constitute an economic demand should reflect the full social cost of supplying the water for this demand. That is, the price consumers pay should include both the direct cost and the cost of internalizing the externalities associated with the supply of water.

The requirement of water users to pay the full social cost for the water they use should be applied to all economic sectors that use water as an input, as well as to households. This requirement may prove particularly important in the case of tourism, as tourism uses abundant quantities of water in semi-arid and arid environments, and is often concentrated around particularly sensitive sites.

The Pricing Implications of Differentiation by use

There are two basic types of water have been identified. The first are the normative needs which should be supplied, regardless of cost or ability to pay. These include human needs, environmental needs, spiritual needs, and the water needed to sustain agriculturally-based peripheral communities with no alternative sources of livelihood within a reasonable commute.It is suggested that these needs can be prioritized, with human needs receiving the top priority, spiritual needs and environmental needs a second priority, and community needs receiving a lower priority. However, as the needs are normatively based, their prioritization is also normative. Hence different people may prioritize the needs differently.

The second type of water is water that should be viewed normatively as a commodity (or factor of production). This water includes water that is used for food production (except in peripheral communities), water used in the industrial, service and tourism sectors, and water for additional domestic use.

The direct implication of this differentiation is that different drops of water entering the same house through the same pipe may be treated differently. While some of the water entering the households is a need, additional water should be viewed as a commodity. This differentiation has direct policy implications.

The water which is viewed as a need should be supplied regardless of the ability to pay. Thus, while it will not necessarily be supplied free of charge, the rates have to be low enough so that they do not prevent the household from utilizing the basic quantity of water that is viewed as a human need. Similarly, water that is used for spiritual needs should also be priced modestly, if at all. In contrast, the water that is viewed as a commodity should in principal be priced at the full social cost. This implies that a block rate pricing system should be adopted, and that the difference between the basic rate applied to needs and the rate applied to commoditized water will be high. All the needs merit subsidization, while additional quantities of water, beyond the needs identified above, should be fully priced.

The difference between the two rates has a spatial dimension. As the cost of supply and externalities vary over space, the upper rates should vary accordingly. In contrast, needs should be provided affordably regardless of spatial variation. Thus the pricing of the first 60 m³/c/yr will not vary over space6. The price of commoditized water, the supply above 60 m³/c/yr, will vary over space.

The pricing of environmental water may pose a particular problem, as there is no direct user. In essence it can be argued that water retained in nature should not be priced at all, as it is not 'supplied'. Indeed in many cases such water is not priced. However, water retained in nature is likely to have a positive shadow price. Hence, it can be argued that it should also be priced. In some countries, such as Israel, this rationale is implemented and nature protection authorities are charged by the water authority for water retained in streams or supplied to sustain ecosystems. But, in such cases, these costs should be subsidized from the general coffer, or cross-subsidized from tourism-based revenues or taxes, where the water bodies are a tourist attraction.

The most difficult policy dilemmas are likely to arise with regard to the provision and pricing of the water necessary to sustain peripheral farming communities. In principle the provision and pricing of water for irrigation should be differentiated according to whether the water is considered a community need, as defined above, or as an input for the production of food in non-peripheral settings, or beyond what is needed to sustain the peripheral agriculture. However, as Molle & Berkoff show in a comprehensive mapping of the issues involved in the pricing of water for irrigation, the pricing of water for irrigation differs substantially from the pricing of water in the domestic sector, both theoretically and practically. Essentially, they argue that pricing is not a useful allocation mechanism under most circumstances and hence has been used primarily to finance the upkeep of irrigation systems. Even in this limited role water is often subsidized. The main policy issues in the agricultural sector are, therefore, the control of distribution and the allocation of water.

In those parts of the world where farming is the basis for community, it is necessary to support irrigation works. This is well recognized, and indeed irrigation is often subsidized. However, the differentiation between community water needs and the water inputs to food production made above cannot be applied only through pricing in this case, as full cost pricing of water for irrigation is unfeasible in most circumstances. Rather, irrigation water policies have to be discussed within the wider scope of farm and social policies. Yet, these are highly contested in most societies, not least because the differentiation between peripheral and non-peripheral farming is far from clear. Moreover, as Moench points out, these relationships are likely to change over time as rural economies diversify. But such diversification is a function of the policies implemented. Hence, the only differentiation that may be feasible is between irrigation by small farmers in remote areas and large-scale industrial food producers, which may leave many if not most of the farmers in the middle between these two groups. Thus, the practical implications of the differentiation between water used for sustaining communities based on peripheral agriculture and water used as an input for food production may have more limited practical policy implications than in the domestic sector.

Equality of all water sources

The discussion so far has largely focused on blue water. But, as was noted at the outset, blue water is but a small fraction of the total water on the planet. With the advent of large-scale desalination and the improvement in wastewater treatment and recycling technologies, a normative discussion of water should be widened to include these additional sources. This raises the question of whether these additional sources of water should be treated in the same manner as blue water.

The main difference between desalinated seawater, and to some extent treated wastewater, and blue water is that the former are essentially humanly produced. In other words, it can be argued that desalinated seawater is an industrial product, and not a natural resource. Indeed, the desalination of seawater alters the basic geography of water, as it does not flow to the sea, but rather from it, and the location of the source as well as the direction and location of the pipes through which they flow are an outcome of political and management decisions.

The same is true, to a lesser extent, with treated wastewater, as the potential for their re-use is a function of the level to which they are treated, which in turn is a function of investments and technology. However, it can be argued that as the freshwater supplied to the domestic sector is treated, and often conveyed over a long distance and in some cases out of the natural basin, this feature of wastewater recycling is not substantially different from blue water. Still, the use of both desalinated seawater and recycled water is likely to differ substantially from blue water.

As desalinated seawater is produced it is inherently commoditized and marketized. Hence, while a government can purchase it and then supply it at a subsidized rate for a basic domestic need, it is likely to be used primarily to supply economic demands, whether in the domestic sector or in the tourism sector. Desalinated seawater, however, is not likely to be seen as a substitute for freshwater in addressing spiritual needs, and is not likely to be used for irrigation, particularly in economically weak peripheral areas. Hence, its only contribution to these needs is by its potential to substitute for fresh water in the domestic sector, thereby allowing more blue water to be used for these needs.

The use of gray water is likely to be constrained by societal concerns to municipal uses and for irrigation. As gray water emanates from the urban centers, the likelihood that it will be used to support peripheral agriculture is small. Hence, its contribution to addressing the various needs is likely to be similar to that of desalinated seawater – by substituting for blue water that may be used for the various needs identified above.

Table below summarizes the relationship between the two types of typologies discussed above. The columns differentiate the waters according to use, while the rows differentiate the waters according to source. The entries in the cells defined by the rows and columns indicate whether the source is likely to supply the use and, if so, what are the policies under which it should be supplied.

As can be seen in Table below, not all combinations of source and use are pertinent. Actually there is only a limited set of waters that comes out of this two-dimensional typology. The first is human needs supplied by blue water. This is the case discussed above, which justifies subsidization.

Table: Waters by type of use and source.

	Normative				Commodity	
Water source	Human	Environmental	Spiritual	Community	Food	Other uses
Blue	S	S	S	Priority allocation	(FC)a	FC
Green	–	+	-	+	+	+
Gray	–	Sb	–	Sc	(FC)a	FC
Produced (desalinated)	Sc	–	–	–	–	FC

Key: +: Water source used but with no policy implications; –: Improbable; S: Supply merits subsidization; FC: Full cost pricing principles should apply;

a: Full-cost pricing is limited to large commercial farming, if at all;

b: Possible, mainly for stream rehabilitation;

c: Possible, but unlikely.

This need can also be satisfied by desalinated seawater. However, as the cost of desalinated seawater is usually substantially higher than freshwater, and as the domestic needs as defined in this paper merit subsidies, the reliance on desalinated seawater to supply domestic needs is unlikely, except in extreme cases such as in arid islands.

The second type of water is for environmental needs, which have to be based on blue water and green water. It is possible, however, to use recycled water in some cases to augment natural flows, particularly as part of river restoration schemes. Yet, this is likely to be of limited scope, as the availability of recyclable wastewater is a function of the existence of upstream human habitation, which is often not the case in natural settings.

The third need, the spiritual need of humans, will invariably be limited to blue water, and may have additional constraints placed on its supply.

These three types of water can be termed as meritorious water, as these merit subsidies and should be supplied regardless of affordability constraints.

The fourth type of water is water required for community needs. These needs are supplied by green water, and where green water is insufficient. It can be augmented by blue water. While the supply of blue water for this need merits support, this support will not be limited to pricing. As most irrigation water is not fully priced for practical and socio-political reasons, the policies in this case have to focus on priorities in water allocations. Water for food production, beyond the community need, should be viewed as an economic input. But with the exception of large-scale commercial farming it is unlikely that full-cost pricing can be applied to it. The water sources for food production are green water, blue water and possibly gray water.

The final type of water is the commoditized and marketized water used to supply other domestic needs, as well as inputs to other economic sectors. While much of this water is supplied by blue water, in closed basin situations in arid and semi-arid situations a growing part of them may be supplied by produced water, mainly through desalination. The policy question that should be at the center of attention with regard to this type of water is whether users are charged the full social cost of water, or whether they are implicitly subsidized at the expense of the environment.

Water has been regularly discussed in the last century as a single substance, which has multiple uses. In this paper I argue that from a normative perspective we should discuss water in the plural term 'waters', as they constitute a variety of 'things' with a similar chemical composition. In this sense it may seem a return to older times, before the modernist unitary view of water became hegemonic. But the meanings of the 'things' that constitute waters today are different from those that were considered in the pre-modernist era. Essentially, I suggest that waters are composed of various needs, those uses with a normative rationale, and 'wants', which constitute an economic demand, and are differentiated also by source – natural, recycled or produced. The needs have also to be differentiated into direct human needs, spiritual needs, environmental needs and community needs, which may be prioritized.

The differentiation by source and use has direct policy implications. In essence, the needs, most of which are to be addressed by blue or green water, should be provided regardless of cost considerations, while the 'wants' should be priced at full social cost, and may be supplied from produced water. Having said this, there are multiple nuances in the policy considerations, some of which have been noted here. Yet a full consideration of all the policy nuances still requires further work.

Regardless of the specific policy ramifications, the main outcome of the shift in view from water to waters is a new differentiation in the way that water should be discussed and analyzed. Rather than referring to domestic, industrial, environmental, agricultural and tourism uses or demands, discussions should refer to needs, wants, natural water, recycled water and produced water. In this paper the basic conceptual framework outlining this new water language is advanced. Such lingual shifts are likely to have widespread ramifications for the ways in which policy problems will be framed in the ever-changing policy arena. But fleshing out the full implications of these changes in the language of water will require further scrutiny and deliberations.

Different Types of Water

Water is one of the key reasons for human survival and civilization in general. Water is considered to be the most important factor behind existence of life on earth. Human body is made of 70 percent of water, much of which is lost though urine and sweat, which is the reason why experts and nutritionists emphasise upon the need to be hydrated at all times. Water carries out many important jobs such as flushing bacteria out of your bladder, aiding digestion, carrying nutrients and oxygen to the cells, preventing constipation, maintaining the electrolyte balance, etc. Water and its importance have been emphasized upon since centuries. In India, traditionally we always offer a glass of water to anyone who comes home, be it a guest or a member of the family returning after a long day at work. The colourless and tasteless elixir of life comes in many types. Here's a water guide for all those wondering what is a sparkling water and how does it differ from the regular mineral varieties.

Types of Water: Water is distinguished on the basis of its origin, consistency, composition and treatment

1. Tap Water

Tap water is the water that you get directly from your faucet, it may or may not be suited for drinking purposes. It is widely used for household chores such as cleaning, cooking, gardening and washing clothes. It must meet the regulations set by the local Municipal bodies.

Tap water is a type of water is the water that you get directly from your faucet

2. Mineral Water

Mineral water is the water that naturally contains minerals

Mineral water is the water that naturally contains minerals. It is obtained from underground sources, which makes it rich in minerals like calcium, magnesium, manganese. No further minerals can be added to the water. The water also cannot be subjected to any treatment, except for limited ones such as carbonation, iron or manganese removal, before packaging. The essential minerals give it a reputation of healthy drinking water. The component of mineral water can vary from brand to brand, some may have more number of minerals while others may have lesser. The presence of the minerals also gives the water a characteristic salty taste.

3. Spring Water

In some places, rainwater accumulated underground tends to "leak" out at the surface as a spring, or puddle. Natural springs are not passed through a community water system and are yet considered suitable for drinking as it comes from under the ground.

4. Well Water

When it rains, water trickles down and travels through the inner crevices of the soil, beneath the ground to form underground lakes. This happens over a period of time. In rural areas, one of the primary sources of water is what is dug out from deep wells. Deep wells directly tap groundwater and bring it to the surface from which people can take their water.

5. Purified Water

A purified water is the water which after deriving from its source has underwent purification treatment in a plant. The act of purifying entails removing all bacteria, contaminants and dissolved solids making it suitable to drinking and other purposes. You can either purchase it from the markets or install a water purifier at home and have pure water to consume.

Type of water which after deriving from its source has underwent
purification treatment is purified water

6. Distilled water

Distilled water or dimineralised water is one where the water has been subjected to a treatment that removes all its minerals and salt by the process of reverse osmosis and distillation. It is an absolutely pure form of water but it is not typically recommended for drinking. It can cause mineral deficiencies because it is devoid of all salts and most of the natural minerals in the water are gone as a result of this process. Drinking this water may cause a rapid sodium, potassium, chloride, and magnesium loss.

7. Sparkling Water

You may have encountered waiters in your favourite restaurant asking your preference of water: regular, mineral or sparkling. Sparkling water is the water that has undergone carbonation which makes your water fizzy just like your sodas. Sparkling water may be spring water, purified water or even mineral water, upon addition of carbon dioxide it becomes sparkling water.

Hydrosphere

The hydrosphere refers to the water on or surrounding the surface of the globe, as distinguished from those of the lithosphere (the solid upper crust of the earth) and the atmosphere (the air surrounding the earth). More specifically, the hydrosphere includes the region that includes all the earth's liquid water, frozen and floating ice, water in the upper layer of soil, and the small amounts of water vapour in the earth's atmosphere. The hydrosphere is the major setting for the earth's hydrologic cycle.

Origin of Water on Earth

The most significant feature of Earth, in contrast to our other neighbouring planets, is the presence of liquid water that covers more than two-thirds of the planet's surface.

This water came about during the early days of the formation of the Earth, when the earth's surface cooled down and the oxygen and hydroxides contained in the accreted material, diffused toward the surface. These gases then cooled and condensed to form the Earth's oceans. It is believed that since then, there has been little loss or gain in the overall quantity of the hydrosphere, despite minor fluctuations like gain from continued degassing and in falling comets; and loss at the upper layers of the atmosphere due to ultraviolet light breaking up of water molecules.

Distribution and Quantity of Water Across the Globe

The earth's water has six major reservoirs in which water resides. These include the oceans, the atmosphere (split into two reservoirs, one over the land and one over the oceans), surface water (including water in lakes, streams, and the water held in the soil), ground water (water held in the pore spaces of rocks below the surface), and snow and ice. The location of some major reservoirs on earth is shown in Figure below,

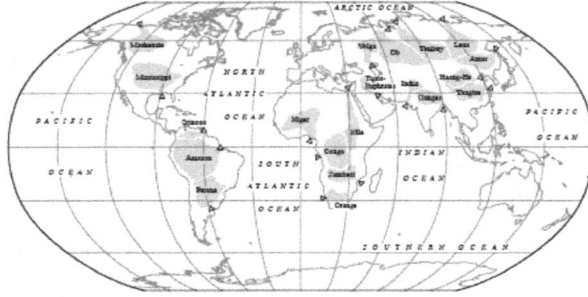

Figure: The location of some major global water reservoirs:
oceans and surface water drainage basins.

The approximate contribution of the different components of the reservoirs to the hydrosphere, the annual recycled volumes and the average replacement periods are shown in Table below.

Table: The distribution of water across the globe.

Location	Volume (10^3 km^3)	% of total volume in hydrosphere	% of freshwater	Volume recycled annually (km^3)	Renewal period (years)
Ocean	1,338,000	96.5	-	505,000	2,500
Ground water (gravity and capillary)	23,400[1]	1.7		16,700	1,400
Predominantly fresh ground water	10,530	0.76	30.1		
Soil moisture	16.5	0.001	0.05	16,500	1
Glaciers and permanent snow cover	24,064	1.74	68.7		
Antarctica	21,600	1.56	61.7		
Greenland	2,340	0.17	6.68	2,477	9,700
Arctic Islands	83.5	0.006	0.24		
Mountainous regions	40.6	0.003	0.12	25	1,600
Ground ice (permafrost)	300	0.022	0.86	30	10,000
Water in lakes	176.4	0.013	-	10,376	17
Fresh	91.0	0.007	0.26		
Salt	85.4	0.006	-		
Marshes and swamps	11.5	0.0008	0.03	2294	5
River water	2.12	0.0002	0.006	43,000	16 days

Biological water	1.12	0.0001	0.003		-
Water in the atmosphere	12.9	0.001	0.04	600,000	8 days
Total volume in the hydrosphere	1,386,000	100	-		
Total Fresh water	35,029.2	2.53	100		

[1] Excluding groundwater in the Antarctic estimated at 2 million km^3, including predominantly freshwater of about 1 million km^3.

Table highlights the enormous disparity between the huge volume of saltwater and the tiny fraction of freshwater and, in addition, the long residence time of polar ice and ground water, as opposed to the brief period for which water remains in the atmosphere. Some 96.5 percent of the total volume of the world's water is estimated to exist in the oceans and only 2.5 percent as freshwater. Of this freshwater, nearly 70 percent is considered to occur in the ice sheets and glaciers in the Antarctic, Greenland and in mountainous areas, while a little less than 30 percent of it is calculated to be stored as groundwater in the world's aquifers.

Water moves through the reservoirs by a variety of processes and at different rates, with unique residence times within any reservoir. This flow of water constitutes the Earth's hydrologic cycle. A brief summary of the major processes involved in this movement, along with the flux, i.e. amount of water transferred per unit time are shown in Figure below.

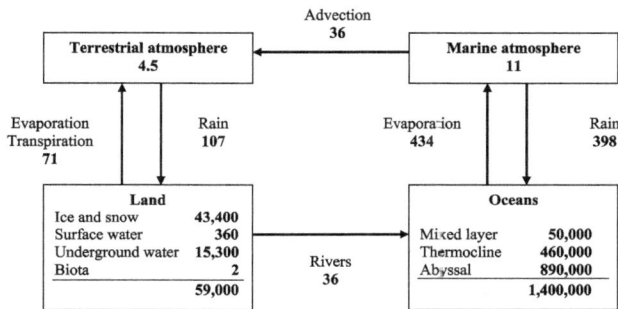

Figure: Estimates of global water reservoirs (in 10^{15} kg and 10^{15} kg yr-1) global water cycle fluxes.

The Biogeochemistry of the Hydrosphere

The quality of natural water in the various reservoirs of the hydrosphere depends on a number of interrelated factors. These factors include geology, climate, topography, biological processes, land use and the time the water has been in residence. Table gives a comparison of major elements present in selected reservoirs.

Table: Chemistry of some hydrosphere components (in parts per million – ppm)

Major Element	average sea water	Average natural river water	Average rain Water
Chloride (Cl⁻)	19,000	5.75	3.79
Sodium (Na⁺)	10,500	5.15	1.98
Sulphate (SO₄²⁻)	2,700	8.25	0.58
Magnesium (Mg²⁺)	1,350	3.35	0.27
Calcium (Ca²⁺)	410	13.4	0.09
Potassium (K)	390	1.3	0.3
Bicarbonate (HCO₃⁻)	142	52	0.12
Bromide (Br²⁺)	67	0.02	
Strontium (Sr)	8	0.03	
Silica (SiO₂)	6.4	10.4	-
Boron (B)	4.5	0.01	
Fluoride (F⁻)	1.3	0.1	

Rainwater has a low concentration of nutrients compared to the other reservoirs. This is because it originates as evaporated water vapour and also has a relatively short residence time in the atmosphere. Even so, it is never pure. The major constituents originate from dissolution of aerosol particles, which are formed from natural process-es, like evaporation of sea spray or human activities, like burning of fossil fuels.

Naturally rain water has a slightly acid pH. This results from the formation of mild car-bonic acid, when rainwater reacts with atmospheric carbon dioxide.

$$CO_2 + H_2O \rightarrow H_2CO_3$$

In areas, with high emission of sulphur dioxide or nitrogen oxide gases from industrial activities or fossil fuel burning, hydrolysis with rain water may result in more acidic rain and pH as low as 4.

River water have an intermediate concentration of ions compared to that of rainwater and oceans. The main factor controlling the composition or river water is the weather-ing reaction between rainfall and rocks that this water passes through. An example is that of calcite in lime stone, which reacts with carbonic acid of rainfall as,

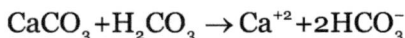

$$CaCO_3 + H_2CO_3 \rightarrow Ca^{+2} + 2HCO_3^-$$

Lakes also have an intermediate concentration when compared to river and sea wa-ter. Lake waters constitute a reservoir of freshwater and their composition is depen-dent upon four factors. These factors being, the hydrology (e.g. relative importance of groundwater or surface water inputs, evaporation); surrounding geology (e.g. car-bonate rocks or granite), temperature-driven circulation patterns, and anthropogenic factors (e.g. acid rain, agricultural fertilizers). In some instances, evaporation of water from lakes formed in closed basins may result in high concentration of salts, as opposed to areas with high rainfall.

Sea and ocean waters are dominated by sodium and chlorine, followed by sulphate and magnesium. Surface sea water is alkaline, with an average pH of about 8. Sea water tends to have a more or less uniform composition in the major elements. But concentrations of minor constituents, including trace and heavy meals and nutrients vary with depth and location, resulting in marked differences in biological productivity. Organisms living on the surface of the sea water are also involved in changes in the composition of sea water, via removal of nutrients and breakdown of organic matter at different depths.

Ground water composition is a result of the rock type it is confined in (e.g. limy is to cal-cium as argillaceous is to silica); and chemical processes of dissolution, hydrolysis, oxi-dation reduction and biological processes. Moreover, anthropogenic contaminants like excess fertilizers and heavy metals may also affect the composition of ground water.

Ice consists a pure solid and has thus only few impurities in its structure. But particu-late matter and gases may be trapped within it. Analysis of successively trapped gases or other anthropogenic substances like carbon dioxide in polar ice caps, has been used to study consecutive changes in atmospheric composition of the past times.

Effect of Human Beings on the Hydrosphere

Over the last 200 or so years, sharp rise in population, urbanization, industrial development and intensification of agricultural practices have combined to largely affect most natural water bodies of the earth. This is due to the transport of the waste products from those activities by surface water, groundwater and the atmosphere. The scale and intensity of this pollution varies considerably, e.g. there are global problems such as presence of heavy metals; regional problems like acid rain; and much more localized ones like groundwater contamination.

Overall, globally, organic material from domestic sewage, municipal waste and agroindustrial effluent is the most widespread pollutant. The sewage contains pathogens which lead to disease and mortality among the populations using this water.

Moreover, this organic material has also high concentrations of nutrients, particularly nitrogen and phosphorus, which cause eutrophication (i.e. nutrient enrichment) of lakes and reservoirs. This eutrophication results in promotion of abnormal plant growth and oxygen depletion, which destroys aquatic ecosystems. Excess fertilizers, from agricultural production areas also have similar consequences.

Acidification of surface waters as a result of acid rain has adverse effects on aquatic life and also human health.

Salinization i.e. the high concentration of salts in the soils of irrigated areas, as a result of poor drainage and high evaporative loss is also a cause of water pollution.

Sediments in the form of suspended load may affect physical structures for e.g. silting up of dams and damage to aquatic life.

Table, shows a summary of the above problems by the type of water bodies polluted and the extent and reach of the effects.

Table: The world's major water quality issues

Issue Scale	Water bodies polluted	Sector affected	Time lag between cause and effect	Effects extent
Organic pollution	Rivers++ Lakes + Groundwater +	Aquatic environment	< 1 year	Local to district
Pathogens	Rivers ++ Lakes + Groundwater +	Health ++	< 1 year	Local
Salinization	Groundwater ++ Rivers +	Most uses Aquatic environment Health	1 - 10 years	District to region
Nitrate	Rivers + Lakes + Groundwater ++	Health	> 10 years	District to region
Heavy metals	All bodies	Health Aquatic environment Ocean fluxes	< 1 to > 10 years	Local to global
Organics	All bodies	Health Aquatic environment Ocean fluxes	1 - 10 years	Local to global

Acidification	Rivers ++ Lakes ++ Groundwater +	health Aquatic environment	> 10 years	District to region
Eutrophication	Lakes ++ Rivers +	Aquatic environment Most uses Ocean fluxes	> 10 years	Local
Sediment load (increase and decrease)	Rivers + Lakes	Aquatic environment Most uses Ocean fluxes	1 – 10 years	Regional
Diversion, dams	Rivers + Lakes + Groundwater ++	Aquatic environment Most uses	1 – 10 years	District to region

+ serious on global scale
++ very serious issue on global scale
WHO/UNEP, 1991

In general, it may be noted that the effect of pollution can reach far beyond the vicinities of its origin. Moreover, the effects may not be noticed until a substantial time has elapsed and it is too late. This calls for constant monitoring and control strategies to save the earth's most precious resource: water.

Fresh Water

Freshwater ecology is a specialized sub category of the overall study of organisms and the environment. Unlike biology, ecology refers to the study of not just organisms but how they react, and are affected by the natural surrounding environment or ecosystem. By studying the plants and animals in a body of water as well as the components of the water itself, a scientist specializing in freshwater ecology can discover vital information about the health and needs of a freshwater system. Freshwater Ecology is a study of the interrelationships between freshwater organisms and their natural and cultural environments.

Classification of Habitats

In studies of the ecology of freshwater rivers, habitats are classified as upland and lowland. Upland habitats are cold, clear, rocky, fast flowing rivers in mountainous areas; lowland habitats are warm, slow flowing rivers found in relatively flat lowland areas, with water that is frequently coloured by sediment and organic matter.

Classifying rivers and streams as upland or lowland is important in freshwater ecology as the two types of river habitat are very different, and usually support very different populations of fish and invertebrate species.

Upland

In freshwater ecology, upland rivers and streams are the fast flowing rivers and streams that drain elevated or mountainous country, often onto broad alluvial plains (where they become lowland rivers). However, altitude is not the sole determinant of whether a river is upland or lowland. Arguably the most important determinants are that of stream power and course gradient. Rivers with a course that drops in altitude rapidly will have faster water flow and higher stream power or "force of water". This in turn produces the other characteristics of an upland river - an incised course, a river bed dominated by bedrock and coarse sediments, a riffle and pool structure and cooler water temperatures.

Figure: Upland lake

Rivers with a course that drops in altitude very slowly will have slower water flow and lower force. This in turn produces the other characteristics of a lowland river - a meandering course lacking rapids, a river bed dominated by fine sediments and higher water temperatures. Lowland rivers tend to carry more suspended sediment and organic matter as well, but some lowland rivers have periods of high water clarity in seasonal low flow periods.

The generally clear, cool, fast-flowing waters and bedrock and coarse sediment beds of upland rivers encourage fish species with limited temperature tolerances, high oxygen needs, strong swimming ability and specialized reproductive strategies to prevent eggs or larvae being swept away. These characteristics also encourage invertebrate species with limited temperature tolerances, high oxygen needs and ecologies revolving around coarse sediments and interstices or "gaps" between those coarse sediments.

Lowland

The generally more turbid, warm, slow-flowing waters and fine sediment beds of lowland rivers encourage fish species with broad temperature tolerances and greater tolerances to low oxygen levels, and life history and breeding strategies adapted to these and other traits of lowland rivers. These characteristics also encourage invertebrate species with broad temperature tolerances and greater tolerances to low oxygen levels and ecologies revolving around fine sediments or alternative habitats such as submerged woody debris ("snags") or submergent macrophytes ("water weed").

Figure: Lowland river

There are four main constituents of the living environment that form the freshwater ecosystem, they are as follows:

- Elements and Compounds of the ecosystem that are absorbed by organisms that are required as a food source or for respiration. Many of these compounds are required by plants and passed along the food chain.

- Plants which are autotrophic by nature, meaning that they synthesize food by harnessing energy from inorganic compounds (plants do so by photosynthesis and the sun); this is done via photosynthesis. These plants (and some bacteria) are the primary producers, as they produce (and introduce) new energy into the ecosystem.

- Consumers, which are the organisms that feed on other organisms as a source of food. These may be primary consumers who feed from the plant material or secondary consumers who feed on the primary consumers.

- Decomposers attain their energy by breaking down dead organic material (detritus), and during this reaction, release critical elements and compounds which in turn are required by plants.

Biotic and Abiotic Factors – Freshwater Ecology

Abiotic factors are essentially non-living components that affect the living organisms of the freshwater community.

When an ecosystem is barren and unoccupied, new organisms colonizing the environment rely on favorable environmental conditions in the area to allow them to successfully live and reproduce. These environmental factors are abiotic factors. When a variety of species are present in such an ecosystem, the consequent actions of these species can affect the lives of fellow species in the area; these factors are deemed biotic factors.

The light from the sun is a major constituent of a freshwater ecosystem, providing light for the primary producers, plants. There are many factors which can affect the intensity and length of time that the ecosystem is exposed to sunlight:

- Aspect: The angle of incidence at which light strikes the surface of the water. During the day when the sun is high in the sky, more light can be absorbed into the water due to the directness of the light. At sunset, light strikes the water surface more acutely, and less water is absorbed. The aspect of the sun during times of the day will vary depending on the time of the year.

- Cloud Cover: The cloud cover of an area will inevitably affect intensity and length of time that light strikes the water of a freshwater ecosystem. Species of plants rely on a critical period of time where they receive light for photosynthesis.

- Season: The 4 seasons in an ecosystem are very different, and this is because less light and heat is available from the sun in Winter and vice versa for Summer, therefore these varying conditions will affect which organisms are suited to them.

- Location: The extreme latitudes receive 6 months of sunlight and 6 months of darkness, while the equator receives roughly 12 hours of sunlight and darkness each day. This sort of variance greatly affects what type of organisms would occupy freshwater ecosystems due to these differences.

- Altitude: For every one thousand metres above sea level, average temperature drops by one degree Celsius. Altitude will also affect the aspect of which sunlight hits the freshwater ecosystem, therefore playing a part on which organisms will occupy it.

Many abiotic factors can play a part in determining the end product, which organisms live and succeed in the freshwater ecosystem. The sun provides light for photosynthesis, but also provides heat giving a suitable temperature for organisms to thrive in. The temperature of a freshwater environment can directly affect the environment as a whole and the organisms that occupy it.

Enzymes operate best at an optimum temperature, and any deviation from this temperature 'norm' will result in below optimum respiration in the organism. All aquatic life are ectotherms, meaning their body temperature varies directly with its environments.

Temperature affects the density of substances, and changes in the density of water means more or less resistance for animals who are travelling in the freshwater environment.

Abiotic Factors - Water Conditions

Evidently, the light and heat from the sun play an important role in providing suitable conditions. However, the water conditions also inevitably have an effect on life in the ecosystem. A still body of water will inevitably be disturbed by various factors, which will affect the distribution of organisms in the water. Wind is considered to be the prime factor responsible for disturbing water, though changes in temperature can create convection currents where temperature is evened out across the body of water via this movement.

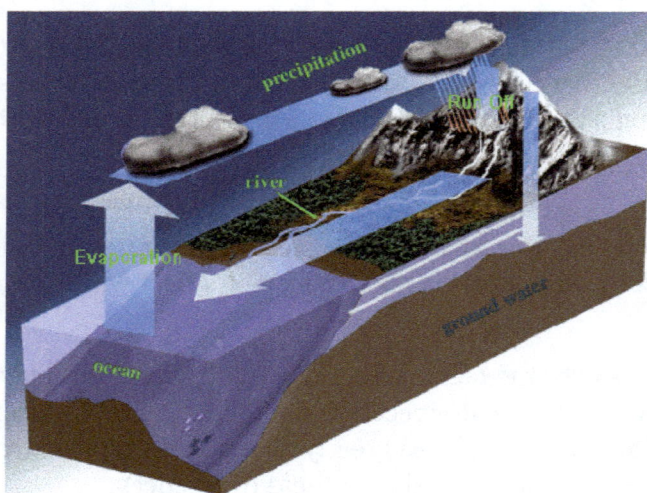

Naturally, a river will have water movement as water succumbs to gravity and moves downstream. These are relatively constant factors that affect water movement though, for example, human intervention can also cause water movement. The surface tension of the water will also affect the organisms that occupy the area, depending on the cohesion of water at the surface; it can affect the amount of oxygen that reaches organisms living below the water surface.

These factors all affect the way of life for organisms occupying such a freshwater ecosystem. On a more molecular level, the chemical compositions of the water, soil and surrounding air also play a part in determining the face of the ecosystem.

The *oxygen* concentration of the water and the surrounding air will have great bearing on which organisms can survive in a particular environment. Oxygen is required for aerobic respiration in animals, and the concentration of oxygen in an area is determined by many factors, including temperature and abundance of organisms for example.

Many chemical reactions and cellular processes rely on the availability of oxygen; therefore the concentration of oxygen in the ecosystem will inevitably alter the ecosystem itself. The same applies to carbon dioxide concentration. CO_2 is required for photosynthesis, and can also affect the pH of the water for example.

The study of ecology in freshwater is usually divided into 2 categories, lentic (still) and lotic (running) water. These two bodies of water also have a bearing on which organisms are likely to occupy the area.

Freshwater Communities and Lentic Waters

Lentic (still water) communities can vary greatly in appearance; anything from a small temporary puddle to a large lake is capable of supporting life to some extent.

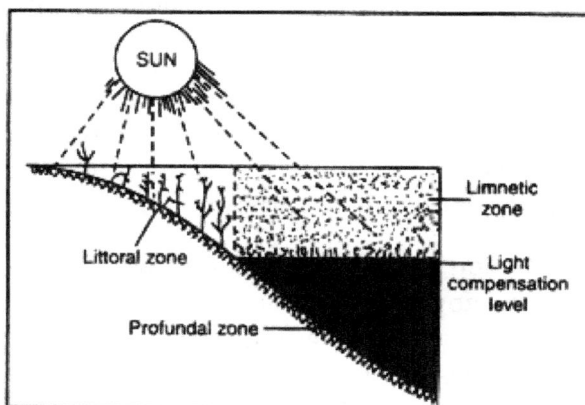

Different zones of Deep Freshwater Lake

The creation of many of today's long standing freshwater lentic environments are a result of geological changes over a long period of time, notably glacial movement, erosion, volcanic activity, and to an extent, human intervention.

The consequence of these actions results in troughs in the landscape where water can accumulate and be sustained over time. The size and depth of a still body of water are major factors in determining the characteristics of that ecosystem, and will continually be altered by some of the causes mentioned above over a long period of time.

One of the important elements of a still water environment is the overall effect that temperature has on it. The heat from the sun takes longer to heat up a body of water as opposed to heating up dry land. This means that temperature changes in the water are more gradual, particularly so in more vast areas of water. When this freshwater ecosystem is habitable, many factors will come into play determining the overall make up of the environment which organisms will have to adapt to.

As with osmosis, temperature will even out across a particular substance over time, and this applies to a still body of water. Sunlight striking the water will heat up the surface, and over time will create a temperature difference between the surface and basin in the body of water. This temperature difference will vary depending on the overall surface area of the water and its depth.

Over time, two distinctly different layers of water become established, separated by a large temperature difference and providing unique ecological niches for organisms. This process is called stratification, where the difference in temperature between surface and water bed are so different they can easily be distinguished apart. The surface area is deemed the epilimnion, which is warmed water as a result of direct contact with sunlight. The lower layer is deemed the hypolimnion, found below the water surface, and due to increased depth, receives less heat from the sun and therefore results in the colder water underneath.

Some factors can affect the amount of light received by autotrophic organisms (organisms that perform photosynthesis) can affect their level of photosynthesis and respiration, hence affect their abundance and therefore and subsequent species that rely on them.

Organic material and sediment can enter the still water environment via dead organisms in the area, and water flowing into the area from hills and streams. Buoyant material will also block out light required by the primary producers of the ecosystem.

When water moves, the friction caused by the moving water against the water bed and its banks will result in disturbing loose sediment. Depending on the weight of this sediment, heavier particles will slowly sink back to the bottom of the body of water while lighter materials will remain suspended in the water. The lightest material will rise to the surface, resulting in less light available to organisms underneath the surface.

Naturally, the consequences of the above will result in less light for organisms that rely on photosynthesis as a means of food, and subsequently means that organisms that feed on these autotrophic organisms will soon find that their food source is less freely available.

Another major factor affecting still water communities is the oxygen concentration of the surrounding area. Oxygen concentration is primarily affected by three factors:

- The surface area which the body of water is exposed to the open air environment.

- The circulation of water, chiefly due to temperature differentiations in different areas of the water body (convection currents).

- Oxygen created as a result of respiration, consumption, and the oxygen consumed by animals and bacteria.

Temperature can also affect the concentration of oxygen available, which in turn, means that the depth of the water will therefore also have an effect. In turn, carbon dioxide levels, which are closely related to the oxygen levels available, will be required by organisms undergoing photosynthesis. The availability of these will affect the organisms in the ecosystem. Their relationships with temperature will also affect their availability. Evidently, some of these factors vary through different conditions, and changes in one of the factors usually results in changes with the others. This is also the case of pH, for example, as an increase in carbon dioxide results in a drop of pH.

Still Water Animals

Through millions of years of evolution, animals living in an aquatic environment have diversified to occupy the ecological niches available in the ecosystem. When studying the habitats of these particular organisms, three main areas of the freshwater environment can be distinctly classified:

- The Profundal Region: An area of still water that receives no sunlight therefore lacks autotrophic creatures. The animals in this zone rely on organic material as a means of food, which is sourced from the more energy rich areas above the profundal region.

- The Pelagic Region: The pelagic region can be found below the surface water, and is defined by the light that is available to it. The pelagic region does not include areas near the shore or sea bed.

- The Benthic Region: The benthic region incorporates all the freshwater environment in contact with land, barring the shallow shore areas. The benthic region is capable of hosting a large volume of organisms, as nutrient and mineral rich sediments are available as a food source while part of the benthic region can occupy the euphotic zone, the area of water where light is available. This will allow an ecological niche for autotrophic organisms which in turn can be a food source for herbivores.

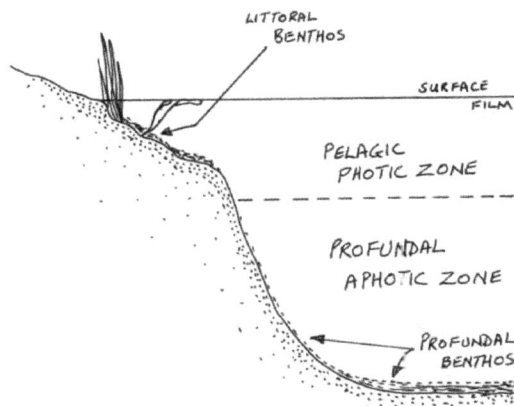

Another distinctive niche for the animal community is that above (epineuston) and below (hyponeuston) the water surface. Epineustic animals receive food from the surrounding hydrosere vegetation, where small animals fall into the water from vegetation and are preyed upon by these epineustic animals.

Below these surface dwelling animals are a collective of animals called the nekton, which live in the pelagic and profundal regions, though rise to the pelagic regions to feed upon these epineustic animals. Fish are included in this nekton community, which play a vital cog in these freshwater communities. Some of these fish are only temporary members of the community, as they move between fresh and salt water. Anadromous fish spawn in freshwater, but live much of their lives in salt water. Catadromous fish are the opposite of this, and spend much of their lives in the freshwater community. Each way, the fish present in the environment at any time form the link between the upper and lower layers of the freshwater community.

Freshwater Lentic Communities and Animals

Plants that live partially or completely submerged in water are deemed hydrophytes. A form of symbiosis occurs with these hydrophyte plants, which provide means for algae and other organisms to survive in the surrounding environment. This is because the hydrophytes provide the conditions for the likes of algae and bacteria to survive in the environment. In return, herbivore animals tend to feed on this rich blanket of algae as opposed to the plants themselves, therefore protecting them from being consumed.

Animals in this environment feed on these algae, and also upon the detritus matter, the organic material that is rich on the water bed. It is an area of abundant organic material because the plants that survive in this area provide a source of food, and also a source of shelter which can provide protection from predators or a location to hatch offspring in a closed protected area.

The ecological niche alongside the still water banks is occupied by plants called hydroseres, which are partially or totally submerged by water along the banks. Some of these hydroseres are rooted in the water, though some of their leaves penetrate the water surface, while others float on the surface, one side in contact with the water, the other side in contact with the open air environment. In essence, hydroseres possess evolutionary adaptations and dithering respiration rates from land plants that have allowed them to adapt in live in such an environment. Such evolutionary adaptation in plants has meant that their physical structure has changed to suit the environment, and therefore making freshwater plants distinctly unique in appearance.

An example of these adaptations is the lack of rigid structures in freshwater plants. This is due to the density of the water (much higher than that of an open air environment), which 'pushes' against the plant in its daily life. This allows such plants to be more flexible against oncoming water tides, and prevents damage to the plant.

As plants require a minimum concentration of gases in their diet such as carbon dioxide, they require a degree of buoyancy so that contact can be made with the open air environment. Adaptations may include:

- Air Spaces: Air spaces in the plant will decrease density and increase buoyancy.

- Broad Leaves: Broader leaves will spread their weight more evenly across the water surface allowing them to float.

- Waxy Cuticle: On the upper half to allow water to run off the surface to prevent the weight of the water dragging the leaves under the surface.

In still water plants, the method of transpiration as a whole is altered in freshwater plants, due to the abundance of water in their external environment, or in the case of some, uptake of water from a wet environment, but loss of water via their leaves in the open air environment.

An example of transpiration problems for such plants is as follows:

- The plant lives in a marshy environment, where roots uptake water from soaked ground, allowing plenty of water to be up taken and transported up and across the plant.

- The difference in water concentration between the plants' leaves and the open air environment is so great that much of the water absorbed is lost to the external environment, meaning the plant loses water rapidly.

- Such a problem is solved by evolutionary adaptations. These adaptations essentially address the issue of re-balancing the critical deviations between the water that is absorbed and lost in a plant.

Freshwater Plants and Nutrients

On top of the need for plants to maintain a suitable water concentration in plant cells, they also require various nutrients which are found in the nutrient rich soil and the surrounding waters. In addition to the carbon, hydrogen and oxygen required for photosynthesis, plants require a range of macro-elements, notably magnesium (Mg), nitrogen (N), phosphorous (P) and potassium (K). Some of these elements, notably the gases, are readily available in the atmosphere, while carbon dioxide is produced from decomposing organic matter. Other elements are readily available in the soil, with nutrients becoming available from decomposing matter adding to the fertility of the surrounding soil. Oxygen becomes available from the photosynthetic activities of plants, which provide the link between oxygen and carbon dioxide concentrations in the area.

Lotic Communities

Running water freshwater communities are also known as lotic communities (lotic meaning running water). Lotic communities are formed by water being introduced to the freshwater body from a variety of sources;

- Rainfall - A percentage of water in the running water community will be present as a result of rainfall directly entering it.

- Ground Surface Water - Deriving from previous rainfall, water will enter the running water community.

- Underground Water - Water absorbed into the soil can also enter.

- Water Table - Deep underground there is a 'water table' which can also provide water for the running water community.

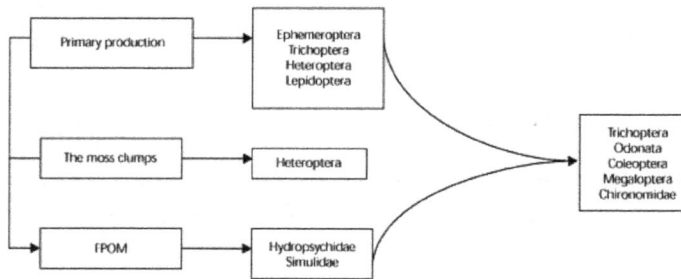

One of the main differences between lotic and lentic communities is the fact that the water is moving at a particular velocity in lotic communities. This can have great bearing on what organisms occupy the ecosystem and what particular ecological niche they can exist in. Running water can bring many factors into play affecting the lives of the organisms in this particular environment:

- Movement of minerals and stones caused by the velocity and volume of the water means the water bed is constantly changing. The faster and higher volume of water present will result in a direct increase in amount and size of particles shifted downstream.

- Standing waves are used by salmon at the bottom of waterfalls to spurn them upstream. At the same time, they cause small air pockets caused by oxygen replacing the splashing water, which results in a small micro-habitat becoming available suited to particular organisms.

- Erosion is caused by the running water breaking down the river bank and beds, causing the geography of the river to change over a long period of time. This means that hydroseres previously occupying the river bank may find themselves distanced from the running water for example, and over time this would mean the overall ecosystem would change over time.

The following is some of the physical and chemical factors that provide the framework of a running water community in which organisms in their favored ecological niches occupy:

- Temperature: The difference between running water and still water temperature is that running water communities' temperature varies more rapidly but over a smaller range. In summer, water from the source of the river is usually colder than the water found at the delta because it has not been exposed to the warm air heated by the sun. The reverse occurs in winter where water is warmer until exposed to the colder air.

- Light: On the whole, less light penetrates a running water body due to ripples in the water, debris blocking out sunlight to lower layers as well as overhanging shrubs that perhaps are taking advantage of a tributary water source. These are all examples of how the intensity of light reaching the lotic community can be affected, and in turn, directly affects the rate of photosynthesis done by plants in the community.

- Chemical Composition: Many factors can alter the chemical composition of the freshwater environment, including precipitation, the percolation of water via vegetation and sea spray to name a few. All in all, various elements and compounds are required by organisms in their daily activities and fluctuations or even an absence of such elements and compounds results in a direct effect on the lives of such organisms.

- Organic Matter: Organic matter previous external to the running water environment can also play a part in altering the ecosystem. This mostly occurs due to overhanging vegetation, although organic matter can be drawn into the ecosystem by the various sources mentioned on the previous page.

Lotic Communities and Algae

In general the diversity of plant species in a lotic community is small compared to that of a still water (lentic) community although small parts of the lotic community host similar conditions to that of a lentic community. Most plants have went through evolutionary adaptations to cope with the force and different conditions that running water brings. Such adaptations have allowed a number of species to successfully take advantage of the lotic community as their ecological niche.

As these conditions are more harsh for a typical species of plant, more notably larger plants, smaller species have found the conditions of the lotic community more favorable. This is due to the fact that they are more flexible in regards to the physical conditions of the water. Algae can grow in all sorts of different places and surfaces, and therefore are a successful constituent of the running water ecosystem. Most of these algae have developed evolutionary adaptations over times that prevent the water current sweeping them away.

There are many species of algae, all of which are capable of growing and reproducing at a quick rate. This consequence results in competition for niches in the freshwater environment, and in light of this, colonies of algae can heavily occupy one area at one moment in time and weeks later they can be succeeded by another species that can succeed in the conditions more favorably.

Algae are also the primary producers of this community, meaning they harness new energy into the ecosystem from the sun which provides the primary consumers with a valuable food source. With this in hand, it is apparent why algae populations and where they can be found in the lotic community is variable on a short-term basis.

Lotic Communities and Animals

The running water environment offers numerous microhabitats that simulate favorable conditions for many types of animals to successfully succeed the freshwater lotic community. As with plants, animals in this ecosystem have also undergone ongoing evolutionary adaptations to better suit this running water environment.

Some of these animals are sessile, meaning they are immobile and fixed to the one place. These animals are usually small, and include the protozoans and some freshwater sponges. These animals either remain attached to the mass of a plant or the water bank surface or rock. They usually obtain their food via tentacles which branch out into the flowing water and form a catchment area that can trap microscopic organisms (such as plankton) that is floating downstream.

As much as these sessile animals have developed adaptations to prevent being washed downstream, they are not thought to be one of the important pillars of the freshwater community. Over time when biotic and abiotic factors affect the landscape of the ecosystem over time, the location of these animals may not be as favorable as it once was, and they are unable to correct this due to their immobile nature. With this in light some animals have developed adaptations that allow them to travel through the water without being inhibited in same spot.

Animals have developed some of the following adaptations over time that helps them cope with the conditions in hand:

- Suckers: These suckers attach themselves to a surface that leeches them into position and can also assist movement in any given direction.

- Hooks or Claws: These sharp objects can dig into any given object and allow the animal to cling to a position or claw their way around the surface.

- Body flattening: This adaptation can allow the animal in the water bear less of the brunt of the force of water moving downstream, therefore reducing it as an inhibitor of their movement. This also allows these animals to enter confined areas (such as under stones) that may present a useful environment for them to live in.

- Streamlining: Just like man-made transport, animals who have underwent streamlining adaptations on their external appearance means that less resistance is presented by the running water when the animal attempts to move.

- Flight: Some animals have adaptations allowing them to fly, removing themselves from the force of the current at ground level and enabling them to move upstream more easily if needs be.

Freshwater Communities and Plankton

Plankton are microscopic organisms that live suspended in the water environment, and form a very important part of the freshwater community. They move via convection or wind induced currents. In almost every habitat of a freshwater ecosystem, thousands of these organisms can be found, and due to their small size and simplicity, they are capable of occupying large expanses of water and multiplying at an exponential rate.

Plankton can be subdivided into two categories:

- Phytoplankton: Phytoplankton are microscopic plants which obtain their energy via photosynthesis. However, some species of bacteria are also capable of photosynthesis and also fall under this taxonomic category. They are important to the ecosystem because they are part of the primary producing community and assist in recycling elements such as carbon and sulphur which are required elsewhere in the community.

- Zooplankton: Zooplankton consist mainly of crustaceans and rotifers, and on the whole are relatively larger than their phytoplankton counterparts. They are relatively unspecialized as their environment does not resist the large populations that can exist in within their environment. Physiologically, there are many evolutionary adaptations that can be found that assist in the buoyancy of them, and prevent their deaths by allowing them to be suspended in the water away from harm.

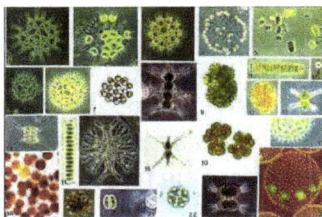

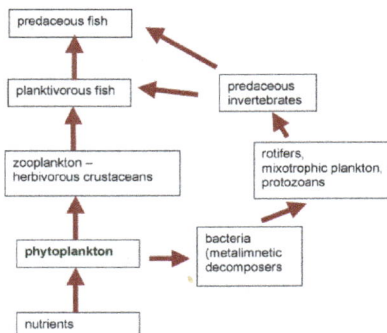

Figure: Food webs in lakes.

Many factors can affect the distribution of plankton in an ecosystem, which has a detrimental effect on the rest of the ecosystem, because as mentioned, they form an essential part of the ecosystem. Phytoplanktons are more abundant in areas with a high intensity of light, as they can convert this light energy into chemical energy while higher temperatures increase growth and multiplication of the both phytoplankton and zooplankton. Elementary, the amount of available nutrients in the environment also plays a part in the distribution and density of phytoplankton.

Pollution in Freshwater Ecosystems

Figure: Fish mortality from water pollution.

As with all ecosystems, the existence and operations of human society inevitably have an effect on the way of life in a freshwater community. Particularly in Western society, where a huge amount of resources are harnessed from the land to fund our lifestyle, there is a resulting effect on the ecosystems of our planet.

- Hot water is used in many industries to cool machinery. This water is removed via a discharge pipe into the river. This increase in temperature can affect the level of oxygen freely available to organisms, which, in turn affects respiration and essentially their way of life. Due to this temperature change, life in the ecosystem is affected.

- Removal of foliage next to a freshwater ecosystem allows more running water to enter its capacity. In light of this, periods of heavy rainfall can result in the water levels fluctuating wildly, which in turn can also affect the temperature of the water quite considerably not to mention all the new chemical agents that would enter the stream from this extra water.

- Recreational use of water bodies such as canoeing also has their effect. Litter from these people can sit on the surface of water and block out sunlight required by the primary producers for photosynthesis. If these primary producers way of life is affected in such a way that their population level decreases, there is a knock on effect to all those organisms who rely on these primary producers for survival.

- At a molecular level, chemicals discharged into the water, notably from industry or pesticides from farmland can affect the freshwater environment considerably. Higher concentrations of particular chemicals (perhaps toxic) mean a lower concentration of essential chemicals required by the organisms of the ecosystem. If this is the case, these organisms cannot perform respiration and function at an optimum level, thus reducing overall biomass in the ecosystem.

Ground Water

Groundwater is fresh water (from rain or melting ice and snow) that soaks into the soil and is stored in the tiny spaces (pores) between rocks and particles of soil. Groundwater accounts for nearly 95 percent of the nation's fresh water resources. It can stay underground for hundreds of thousands of years, or it can come to the surface and help fill rivers, streams, lakes, ponds, and wetlands. Groundwater can also come to the surface as a spring or be pumped from a well. Both of these are common ways we get groundwater to drink. About 50 percent of our municipal, domestic, and agricultural water supply is groundwater.

Process of storage of the ground water

Groundwater is stored in the tiny open spaces between rock and sand, soil, and gravel. How well loosely arranged rock (such as sand and gravel) holds water depends on the size of the rock particles. Layers of loosely arranged particles of uniform size (such as sand) tend to hold more water than layers of rock with materials of different sizes. This is because smaller rock materials settle in the spaces between larger rock materials, decreasing the amount of open space that can hold water. Porosity (how well rock material holds water) is also affected by the shape of rock particles. Round particles will pack more tightly than particles with sharp edges. Material with angular-shaped edges has more open space and can hold more water.

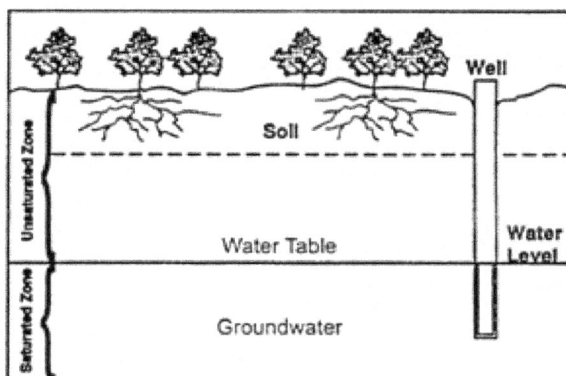

Figure: Groundwater Zones

Groundwater is found in two zones. The unsaturated zone, immediately below the land surface, contains water and air in the open spaces, or pores. The saturated zone, a zone in which all the pores and rock fractures are filled with water, underlies the unsaturated zone. The top of the saturated zone is called the water table. The water table may be just below or hundreds of feet below the land surface.

Aquifer

Where groundwater can move rapidly, such as through gravel and sandy deposits, an aquifer can form. In an aquifer, there is enough groundwater that it can be pumped to the surface and used for drinking water, irrigation, industry, or other uses.

For water to move through underground rock, pores or fractures in the rock must be connected. If rocks have good connections between pores or fractures and water can move freely through them, we say that the rock is permeable. Permeability refers to how well a material transmits water. If the pores or fractures are not connected, the rock material cannot produce water and is therefore not considered an aquifer. The amount of water an aquifer can hold depends on the volume of the underground rock materials and the size and number of pores and fractures that can fill with water.

An aquifer may be a few feet to several thousand feet thick, and less than a square mile or hundreds of thousands of square miles in area. For example, the High Plains Aquifer underlies about 280,000 square miles in 8 states— Colorado, Kansas, Nebraska, New Mexico, Oklahoma, South Dakota, Texas, and Wyoming.

Filling of Water in an Aquifer

Aquifers get water from precipitation (rain and snow) that filters through the unsaturated zone. Aquifers can also receive water from surface waters like lakes and rivers. When the aquifer is full, and the water table meets the surface of the ground, water stored in the aquifer can appear at the land surface as a spring or seep. Recharge areas are where aquifers take in water; discharge areas are where groundwater flows to the land surface. Water moves from higher-elevation areas of recharge to lower-elevation areas of discharge through the saturated zone.

Circulation of Water

Surface water and groundwater are part of the hydrologic cycle, the constant movement of water above, on, and below the earth's surface. The cycle has no beginning and no end, but you can understand it best by tracing it from precipitation. Precipitation occurs in several forms, including rain, snow, and hail. Rain, for example, wets the ground surface. As more rain falls, water begins to filter into the ground. How fast water soaks into, or infiltrates the soil depends on soil type, land use, and the intensity and length of the storm. Water infiltrates faster into soils that are mostly sand than into those that are mostly clay or silt. Almost no water filters into paved areas. Rain that cannot be absorbed into the ground collects on the surface, forming runoff streams.

When the soil is completely saturated, additional water moves slowly down through the unsaturated zone to the saturated zone, replenishing or recharging the groundwater. Water then moves through the saturated zone to groundwater discharge areas.

Evaporation occurs when water from such surfaces as oceans, rivers, and ice is converted to vapor. Evaporation, together with transpiration from plants, rises above the Earth's surface, condenses, and forms clouds. Water from both runoff and from groundwater discharge moves toward streams and rivers and may eventually reach the ocean. Oceans are the largest surface water bodies that contribute to evaporation.

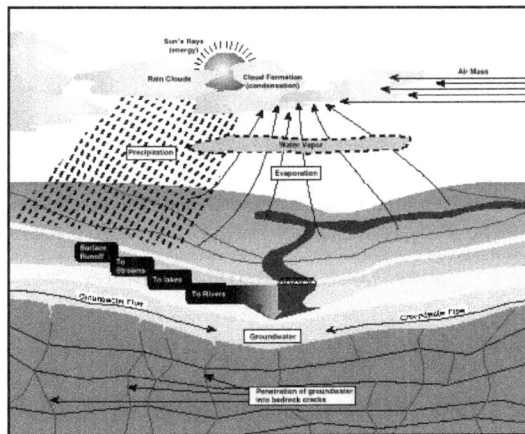

Figure: Hydrologic Cycle

Contamination of Groundwater

Groundwater can become contaminated in many ways. If surface water that recharges an aquifer is polluted, the groundwater will also become contaminated. Contaminated groundwater can then affect the quality of surface water at discharge areas. Groundwater can also become contaminated when liquid hazardous substances soak down through the soil into groundwater.

Contaminants that can dissolve in groundwater will move along with the water, potentially to wells used for drinking water. If there is a continuous source of contamination entering moving groundwater, an area of contaminated groundwater, called a plume, can form a combination of moving groundwater and a continuous source of contamination can, therefore, pollute very large volumes and areas of groundwater. Some plumes at Superfund sites are several miles long. More than 88 percent of current Superfund sites have some groundwater contamination.

Groundwater Contamination by Liquids

Some hazardous substances dissolve very slowly in water. When these substances seep into groundwater faster than they can dissolve, some of the contaminants will stay in liquid form. If the liquid is less dense than water, it will float on top of the water table, like oil on water. Pollutants in this form are called light non-aqueous phase liquids (LNAPLs). If the liquid is more dense than water, the pollutants are called dense non-aqueous phase liquids (DNAPLs). DNAPLs sink to form pools at the bottom of an aquifer. These pools continue to contaminate the aquifer as they slowly dissolve and are carried away by moving groundwater. As DNAPLs flow downward through an aquifer, tiny globs of liquid become trapped in the spaces between soil particles. This form of groundwater contamination is called residual contamination.

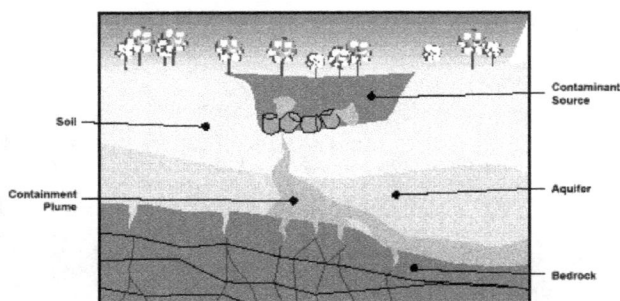

Figure: Contaminated Groundwater

Factors Affecting Groundwater Contamination

Many processes can affect how contamination spreads and what happens to it in the groundwater, potentially making the contaminant more or less harmful, or toxic. Some of the most important processes affecting hazardous substances in groundwater are advection, sorption, and biological degradation.

- Advection occurs when contaminants move with the groundwater. This is the main form of contaminant migration in groundwater.

- Sorption occurs when contaminants attach themselves to soil particles. Sorption slows the movement of contaminants in groundwater, but also makes it harder to clean up contamination.

- Biological degradation happens when microorganisms, such as bacteria and fungi, use hazardous substances as a food and energy source. In the process, contaminants break down and hazardous substances often become less harmful.

Difficulties in Cleaning up Groundwater

Cleaning up contaminated groundwater often takes longer than expected because groundwater systems are complicated and the contaminants are invisible to the naked eye. This makes it more difficult to find contaminants and to design a treatment system that either destroys the contaminants in the ground or takes them to the surface for cleanup. Groundwater contamination is the reason for most of Superfund's long-term cleanup actions. Diagram illustrates groundwater treatment in action.

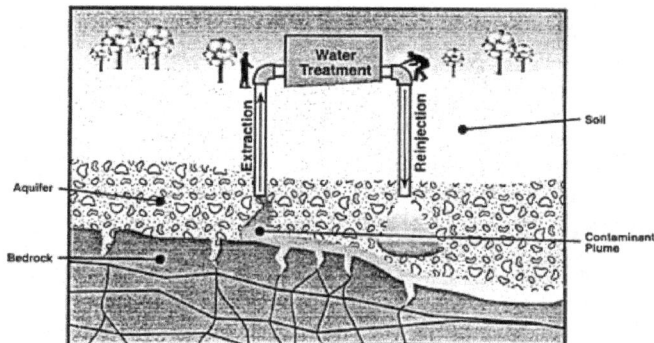

Figure: Pumping and Treating Contaminated Groundwater

Polar Ice Caps

The polar ice caps cover the territory around the north and south poles of Earth, including almost the entire continent of Antarctica, the Arctic Ocean, most of Greenland, parts of northern Canada, and bits of Siberia and Scandinavia. The ice at the North Pole floats on the ocean in the form of a relatively thin sheet. The Greenland and Antarctic ice caps are dome-shaped sheets of ice that feed ice to other glacial formations, such as ice sheets, ice fields, and ice islands. They remain frozen year-round, and they serve as sources for glaciers that feed ice into the polar seas in the form of icebergs. Because the polar ice caps are very cold (temperatures in Antarctica have been measured 126.8° F [−88° C]) and exist for a long time, the caps serve as deep-freezes for geologic information that can be studied by scientists. Ice cores drawn from these regions contain important data for both geologists and environmental scientists about paleoclimatology (prehistoric climate variations) and give clues about the effects human activities are currently having on the world.

Polar ice caps also serve as reservoirs for huge amounts of Earth's water. Hydrologists suggest that three-quarters of the world's freshwater is locked up in the ice sitting on Greenland and Antarctica. The Antarctic ice cap alone contains over 90% of the world's freshwater ice, some in huge sheets over 2.5 mi (4 km) deep and averaging 1.5 mi (2.4 km) deep across the continent. It has been estimated that enough water is locked up in Antarctica to raise sea levels around the globe over 240 ft (73 m).

Although the polar ice caps have been in existence for millions of years, scientists disagree over exactly how long they have survived in their present form. It is generally agreed that the polar cap north of the Arctic Circle, which covers the Arctic Ocean, has undergone contraction and expansion through some 26 different glaciations in just the past few million years. Parts of the arctic have been covered by the polar ice cap for at least the last five million years, with estimates ranging up to 15 million. The Antarctic ice cap is more controversial; although many scientists believe extensive ice has existed there for 15 million years, others suggest that volcanic activity on the western half of the continent it covers causes the ice to decay, and the current south polar ice cap is therefore no more than about three million years old.

At least five times since the formation of Earth, because of changes in global climate, the polar ice has expanded north and south toward the equator and has stayed there for at least a million years. The earliest of these known ice ages was some two billion years ago, during the Huronian epoch of the Precambrian era. The most recent ice age began about 1.7 million years in the Pleistocene epoch. It was characterized by a number of fluctuations in North polar ice, some of which expanded over much of modern North America and Europe, covered up to half of the existing continents, and measured as much as 1.8 mi (3 km) deep in some places. These glacial expansions locked up even more water, dropping sea levels worldwide by more than 300 ft. Animal species that had adapted to cold weather, like the mammoth, thrived in the polar conditions of the Pleistocene glaciations, and their ranges stretched south into what is now the southern United States.

The glaciers completed their retreat and settled in their present positions about 10,000 to 12,000 years ago. There have been other fluctuations in global temperatures on a smaller scale, however, that have sometimes been known popularly as ice ages. The 400-year period between the fourteenth and the eighteenth centuries is sometimes called the Little Ice Age. Contemporaries noted that the Baltic Sea froze over twice in the first decade of the 1300s. Temperatures in Europe fell enough to shorten the growing season, and the production of grain in Scandinavia dropped precipitously as a result. The Norse communities in Greenland could no longer be maintained and were abandoned by the end of the fifteenth century. Scientists argue that data indicate we are currently in an interglacial period, and that North polar ice will again move south some time in the next 23,000 years.

Investigation of Polar Ice Caps

Scientists believe the growth of polar ice caps can be triggered by a combination of several global climactic factors. The major element is a small drop (perhaps no more than 15° F [9° C]) in average world temperatures. The factors that cause this drop can be very complex. They include reductions in incoming solar radiation, reflection of that energy back into space, fluctuations in atmospheric and oceanic carbon dioxide and methane levels, increased amounts of dust in the atmosphere such as that resulting from volcanic activity, heightened winds—especially in equatorial areas—and changes in thermohaline circulation of the ocean. The Milankovitch theory of glacial cycles also cites as factors small variations in Earth's orbital path around the sun, which in the long term could influence the expansion and contraction of the polar ice caps. Computer models based on the Milankovitch theory correlate fairly closely with observed behavior of glaciation over the past 600 million years.

Scientists use material preserved in the polar ice caps to chart these changes in global glaciation. By measuring the relationship of different oxygen isotopes preserved in ice cores, they have determined both the mean temperature and the amount of dust in the atmosphere in these latitudes during the recent ice ages. Single events, such as volcanic eruptions and variations in solar activity and sea level, are also recorded in polar ice. These records are valuable not only for the information they provide about past glacial periods; they serve as a standard to compare against the records of more modern periods. Detailed examination of ice cores from the polar regions has shown that the rate of change in Earth's climate may be much greater that previously thought. The data reflect large climatic changes occurring in periods of less than a decade during previous glacial cycles.

Scientists also use satellites to study the thickness and movements of the polar ice caps. Information is collected through radar, microwave, and even laser instruments mounted on a number of orbiting satellites. Scientists have also utilized similar technology to confirm the existence of polar ice caps on the moon and Mars. These relict accumulations are indicative of the history of these bodies and may prove useful in future exploration efforts as a water and fuel source. The detailed and frequent observations provided by the space-based tools permit scientists to monitor changes in the ice caps to a degree not possible by previous land-based methods.

Recent findings suggest that the ice sheets may be changing much more rapidly than previously suspected. Portions of the ice sheets in Greenland, west Antarctica, and the Antarctic Peninsula are rapidly thinning and, more importantly, losing mass. Scientists are able to document modifications of ice accumulations rates, volume of melt water, and the impact of elevated sea-water temperature and utilize this information in characterizing the movement and evolution of these ice sheets. Glaciers flowing to the ocean in these areas.

2

Properties of Water

Water is a polar inorganic compound, which is tasteless and odorless at room temperature. It is the most abundant substance found on Earth in the form of solid, liquid and gas. The principal properties of water like taste and odor, color and appearance, polarity and bonding, etc. have been carefully analyzed in this chapter.

Salinity

Rising salinity levels in water resources and dryland salinity are the biggest environmental problems facing Western Australia. Salinity is also affecting our natural environment, rural towns, heritage buildings and infrastructure such as roads and railways. This Water Facts sheet explains why stream salinity has increased, the impacts on our surface water resources, and what is being done to tackle the problem.

Salinisation

Most people are aware that the crusty white deposit they sometimes see on bare patches of cleared agricultural land is salt. This is a visible sign of increasing salinity in our landscape, and is sometimes called 'dryland salinity'. Less visible, but still a huge problem, are rising salinity levels in our water resources. Some lakes are naturally saline (for example, the salt lakes in the eastern Avon catchment) but where previously fresh lakes have become salty, the dead vegetation and salt encrustations indicate rising salinity levels. Increasing salt levels are being observed in many south west rivers.

Figure: Salinity in the Avon catchment.

Salt in the Soil and Groundwater

Salt is a problem when it dissolves in rising groundwater and comes to the ground surface or flows into streams. But a huge amount of salt is stored deep in our soils, and as long as it stays there, it poses no problem.

Western Australia's soils have accumulated massive amounts of salt brought in by rain from the sea over tens to hundreds of thousands of years. Salt has been deposited on south-west Western Australia at a rate of 20 to 200 kilograms per hectare per year. Depending on the location and soil type, soils now store between 300 and 10 000 tonnes of salt per hectare deep in the soil profile. In the wheatbelt, for example, the soil holds about 3000 tonnes of salt per hectare.

Rising Watertables Bring Salt to the Surface

Groundwater is water that occupies the pores or crevices in soil or rock.

Stream salinisation happens in two ways:

(i) When rising groundwater carries salt to the land surface where it can be washed away into streams and lakes.

Salinity problems arise when the watertable (the level at which the soil is saturated with groundwater) rises, dissolving salt stored in the soil profile as it does so, and bringing very salty water to or near the ground surface.

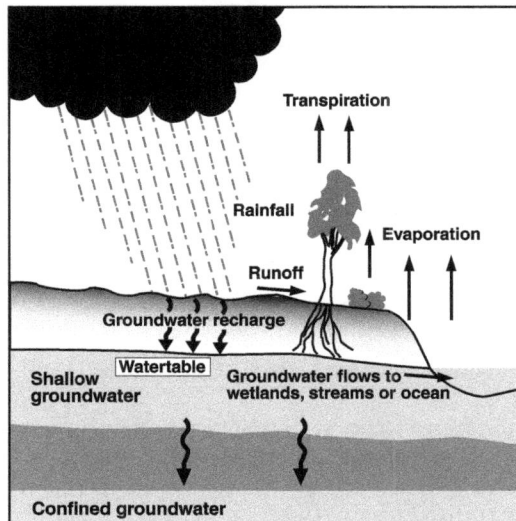

Figure: The water cycle.

In nature, trees play an important role as pumps. Their roots reach deep into the soil and draw water and nutrients up into the leaves. Water is being constantly released from the leaves by evaporation and transpiration. The natural vegetation in the south west is adapted to using a high proportion of the rainfall, so only a small amount recharges the groundwater. A typical large tree in a healthy jarrah forest, in a high rainfall catchment could use an average of 50 litres a day all year round.

The millions of leaves in a forest or woodland also catch a lot of rain. Much of this water evaporates without ever reaching the ground.

When the trees are cleared and replaced with shallowrooted annual crops and pastures, less water is drawn from the ground, and more water falls on the ground during rain. After clearing, the volume of water soaking into the ground may increase more than 10 times.

The result is a rise in the watertable. Watertable rises of 20 to 150 centimetres per year have been measured after clearing. The landscape fills up with water. The water also dissolves salt stored in the soil, and brings it up to the surface.

This is called dryland salinity. This salt can then be washed away by surface flows and carried into streams and lakes.

In irrigated areas, excess irrigation and leakage from irrigation channels can also contribute to the problem.

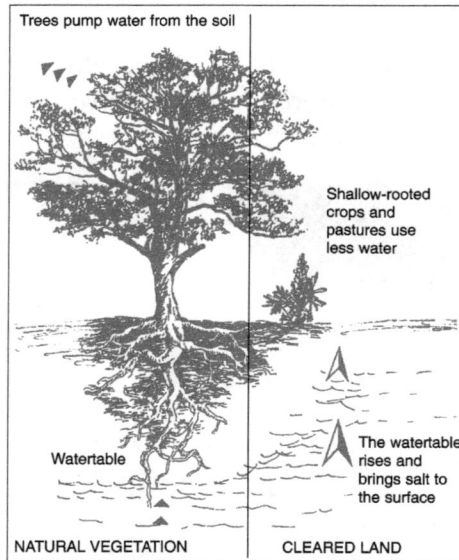

Figure: The rising watertable brings salt from deep in the soil up to the surface where it kills plants, makes water supplies salty, and raises salt levels in rivers.

(ii) When increasing groundwater flows carry larger quantities of salt directly to streams.

With rising watertables, groundwater seepage into streams increases, and greater quantities of salt are dissolved and may be discharged directly into the stream. Salt is then carried downstream to fresher parts of the river.

Both these processes account for a steady increase in salinity of most south-west rivers. For example salinity in the Warren River, whose catchment is 40% cleared, has gone from less than 300 mg/L in 1940 to about 800 mg/L in 1985. The Blackwood River, (catchment 85% cleared) has risen from about 500 mg/L in the mid-1950s to nearly 2000 mg/L in the 1980s.

Impacts of Salinity

The highly saline water that comes up to the land surface kills natural vegetation and crops, and damages soil structure, buildings and roads. Water becomes unsuitable for stock and on-farm water supplies. Stream salinity starts to rise. This may start to happen years before salt on the surface is visible.

Streams and rivers that are fed by saline runoff and groundwater are gradually degraded with serious effects on the plant and animal life. Fringing vegetation is replaced by more salt-tolerant species or introduced grasses. The water is undrinkable and unusable for irrigation. The effects can extend many kilometres down stream from the source of the saline water. Rising watertables and increasing groundwater discharge may also cause inundation and waterlogging of streamlines or low lying land. This, together with the increasing salinity, has probably contributed to the stands of dead and dying trees along many inland watercourses.

The Extent of the Salinity Problem

Western Australia is badly affected, having 70% of Australia's dryland salinity-affected land. In the south west, 18 million hectares of the 25 million hectares originally covered by native vegetation have been cleared. Of this, about 10% (1.8 million hectares) is now salt affected to some degree. Salinity has significantly affected over 80% of waterways in the south west including our divertible water resources (i.e. surface water that has potential for domestic or commercial supply).

Water Supplies

Water supplies are threatened, with over a third (36%) of divertible water resources brackish or saline and 16% marginally saline. Many millions of dollars have already been spent on alternative water supplies to replace storages lost to salinity, for example Harris Dam. Industry suffers because there are costs associated with using higher salinity water and mechanical equipment is likely to have a shorter life.

Rivers and Wetlands

Most of our major rivers in the south west have high increases in salinity each year. Increasing salinity in rivers and wetlands has caused changes to the plants and wildlife. The overall impact is loss of habitat, biodiversity and recreational assets.

Animals that rely on fresh water, such as frogs and water birds are particularly affected.

Frogs are the most sensitive to changes in salinity, and tadpoles can be indicators of salinity changes along streamlines. The slender tree frog is likely to become extinct in inland agricultural areas, and the long-necked turtle may also be at risk from salinisation.

Tackling Salinity

Restoring the Water Balance

The key to controlling the salinity problem is to get the water balance back into equilibrium. That means using more water, lowering the watertable and getting the salt back down to where it is harmless. The main way to do this is to plant trees and deep-rooted perennial crops to take water from the soil. This is a huge task. It has been estimated that $3 billion will be needed over the next 30 years to fix the problem, and over 3 million hectares of appropriate trees and shrubs will have to be planted.

The Salinity Strategy

The Western Australian Government has adopted a comprehensive Salinity Program, with an Action Plan first released in 1996 and updated with public review in 2000. The strategy is a framework for Government, farmers and the community to work together to:

- Reduce further deterioration of agricultural land.
- Recover or rehabilitate salt-affected land.
- Protect natural values (biological diversity).
- Protect water resources.
- Protect infrastructure.
- Give land managers the capacity to address salinity issues.

Key activities include:

- Clearing controls (and in some places clearing bans) to protect vegetation.
- Protecting remnant vegetation under the Remnant Vegetation Protection Scheme which helps to provide funding for fencing.
- Research on the nature and extent of salinity.
- Improved management of crops and pastures to use more water.
- Introduction of tree crops into farming systems.
- Revegetation of cleared land through a large-scale tree planting program.

Water Resource Recovery Catchments

Major rivers currently supplying public water, or having the potential to be used for this purpose in the future, are affected by salinity. The Helena, Collie and Denmark rivers contribute to existing water supply reservoirs and the Kent and Warren rivers are considered to be potential future public water sources.

Concern about salinity led to statutory control of clearing in catchments of these rivers during the late 1970s. The State Salinity Strategy identifies these catchments as requiring priority management. They are termed Recovery Catchments. Recovery Teams of local community and government representatives have been formed. They are working on catchment planning based on salinity risk assessments, and using 'best practices' in land management. The aim is to achieve potable water quality in each of the Recovery Catchment rivers within 20-30 years.

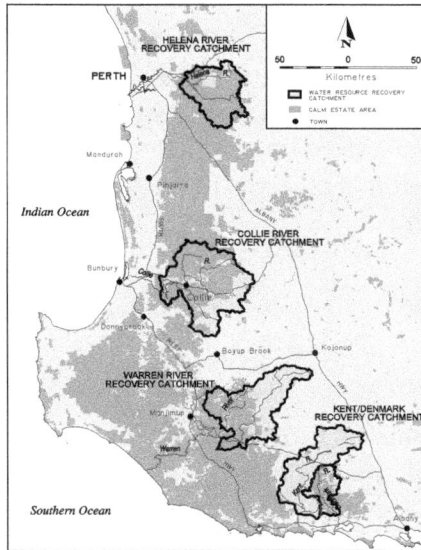

Recovery catchments

Warren and Collie River Recovery Team members visit a tree trial established by CALM.

Tasteless

Taste may be the most difficult of the five senses to pinpoint because a flavor cannot be quantitated; flavor is a sensation that is the sum of smell and taste. The flavor that everyday drinking water produces can be attributed to the minerals and elements that water absorbs as the universal solvent. A typical, commendable drinking water will typically have traces of Potassium, magnesium, calcium and even small amounts of sodium give water its fullness potassium, calcium, magnesium, silica and even some sodium in order to give water a full flavor that is neither dull nor flat. "Good" water cannot be a water that is totally distilled of any impurities. If distilled water tasted the best, it would be the most commercially sold water.

Additionally, recent research suggests that pure H_2O may produce a flavor that cannot be created by lingering minerals in water. At the beginning of the current century, a team at the University of Utah discovered that the taste cells in mammals create proteins called aquaporins, which serve to channel water through cell membranes. The aquaporins provide a possible way for water to stimulate taste cells directly. Currently, the taste buds that are designated to be stimulated by pure water have not been found. It will take further investigation of all areas of the tongue to precisely identify the region of taste buds that taste water.

Odorless

Pure water (distilled) is odorless. But, the drinking water sometimes have this smell due to chlorine and other disinfectants. Also, some impure water having certain impurities or biological waste makes it have odor (unpleasant one). Rainwater is "almost" distilled water, hence won't smell. The smell we experience during rainfall, is when the rainwater mixes with the environment, soil (that a pleasant one!), then we experience odor.

An unwanted odor coming from drinking water is often the sign of greater issues afflicting the tap or pipelines. Often it does not signify the presence of a harmful contaminant, but it could indicate that the distribution system requires flushing or other maintenance. Sometimes water may taste like metal or have an odor like chlorine or bleach.

Bleach Odor

Because public water sources are treated with chlorine, drinking water can sometimes exude an odor similar to bleach or a swimming pool when water is over chlorinated. Small amounts of chlorine are added to public water sources and treatment plants to eliminate traces of bacteria, viruses and parasites as the water travels to its point of use.

The U.S. Environmental Protection Agency requires that chlorine levels in public water systems be maintained at a range that is detectable but not above 4 milligrams per liter (mg/L). Most people will detect a bleach smell if the chlorine level is around 1 mg/L. If the bleach or chlorine odor is strong, the local water source could add extra chlorine to keep the water clean for longer lengths of time to cover longer distances.

The strength of this chlorine scent will usually depend on the distance of a public water source from a household. However, the odor can also be affected by the temperature of the water because colder water can hold on to chlorine for longer. Exposures to levels below 4 mg/L do not pose a risk to your health, but it is relatively easy to eliminate the smell with time or a carbon water filter. Chilling it in the refrigerator or adding a slice of lemon or orange will remove the bleach odor from the water. Many water suppliers use chloramine residuals in their distribution systems. Chloramines have less taste and odor than free chlorine at similar concentrations.

Musty Odor

While generally safe to use and drink, water with a musty odor can in rare circumstances be attributed to pollution. Algal blooms in the source water often contribute a musty taste. For users of city water supplies a musty smell often results from sediment leaching into the plumbing system. Over time, decaying organic matter in the source can produce earthy aromas. Similarly, severely corrosive water in pipes may cause trace amounts of copper, iron or even lead to appear in water, adversely affecting its taste.

Color of Water

It may be true that a bit of color in water may not make it harmful to drink. but it certainly makes it unappealing to drink. So, color in our water does matter when it comes to drinking it, as well as in water for other home uses, industrial uses, and in some aquatic environments.

Pure Water and Color

Is pure water really clear? First, you will rarely see pure water as it is not found in a natural setting. The everyday water you see contains dissolved minerals and often suspended materials. But, for practical purposes, if you fill a glass from your faucet the water will look colorless to you. The water is in fact not colorless; even pure water is not colorless, but has a slight blue tint to it, best seen when looking through a long column of water. The blueness in water is not caused by the scattering of light, which is responsible for the sky being blue. Rather, water blueness comes from the water molecules absorbing the red end of the spectrum of visible light. To be even more detailed, the absorption of light in water is due to the way the atoms vibrate and absorb different wavelengths of light.

Color and Drinking Water

Iron stains can develop over time.

If you have ever drunk water containing a bit of iron in it, you would know from the metallic taste left in your mouth that dissolved chemicals in drinking water can be less than desirable. Color in drinking water can be caused by dissolved and suspended materials, and a brown shade in water often comes from rust in the water pipes. Although water can contain contaminants, which are usually removed by water-supply systems, the plus side is that the water you drink likely contains a number of dissolved minerals that are beneficial for human health. And, if you have ever drunk "pure" water, such as distilled or deioninzed water, you would have noticed that it tasted "flat". Most people prefer water with dissolved minerals, although they still want it to be clear.

Have you ever gotten a glass of water from your faucet and the water is milky white water or hazy? This is almost always caused by air in the water. To see if the white color in the water is due to air, fill a clear glass with water and set it on the counter. Observe the glass of water for 2 or 3 minutes. If the white color is due to air, the water will begin to clear at the bottom of the glass first and then gradually will clear all the way to the top. This is a natural phenomenon and is caused by dissolved air in the water that is released when the faucet is opened. When you relieve the pressure by opening the faucet and filling your glass with water, the air is now free to escape from the water, giving it a milky appearance for a few minutes.

Color and Water in the Environment

Color in water you see around you can be imparted in two ways: dissolved and suspended components. An example of dissolved substances is tannin, which is caused by organic matter coming from leaves, roots, and plant remains (left-side picture). Another example would be the cup of hot tea your grandmother has in the afternoon. In the picture below the color is probably attributable to naturally dissolved organic acids formed when plant material is slowly broken down by into tiny particles that are essentially dissolved in the water. If you filtered that tannin-water in the picture the color would probably remain.

Color caused by dissolved matter: tannis

color caused by suspended material: sediment

Most of the color in water you see around you comes from suspended material, as you can see in the right-side picture of a tributary contributing highly-turbid water containing suspended sediment (fine particles of clay, since this picture is in Georgia) to clearer, but still colored, water in the main stem of the river. Algae and suspended sediment particles are very common particulate matter that cause natural waters to become colored. Even though the muddy water below would not be appealing to swim in, in a way that water has less color than the water containing dissolved tannins. That is because suspended matter can be filtered out of even very dirty-looking water. If the water is put into a glass and left to settle for a number of days, most of the material will settle to the bottom (this method is used in sewage-treatment facilities) and the water will become clearer and have less color. So, if an industry wanted needed some color-free water for an industrial process, they would probably rather start with the sediment-laden water, rather than the tannin colored water.

Suspended material in water bodies may be a result of natural causes and/or human activity. Transparent water with a low accumulation of dissolved materials appears blue. Dissolved organic matter, such as humus, peat or decaying plant matter, can produce a yellow or brown color. Some algae or dinoflagellates produce reddish or deep yellow waters. Water rich in phytoplankton and other algae usually appears green. Soil runoff produces a variety of yellow, red, brown and gray colors.

Effects of Color on Ecosystems

Highly colored water has significant effects on aquatic plants and algal growth. Light is very critical for the growth of aquatic plants and colored water can limit the penetration of light. Thus a highly colored body of water could not sustain aquatic life which could lead to the long term impairment of the ecosystem. Very high algal growth that stays suspended in a water body can almost totally block light penetration as well as use up the dissolved oxygen in the water body, causing a eutrophic condition that can drastically reduce all life in the water body. At home, colored water may stain textile and fixtures and can cause permanent damage.

Colored Dissolved Organic Matter

Colored or Chromophoric Dissolved Organic Matter:

(CDOM) is an important optical constituent in water often dominating absorption in the blue. It is based on the absorption or fluorescence by material passing through a given filter (most often with pore size of $0.2 \mu m$). As such, it is an absorption (or fluorescence)-weighted sum of the different dissolved materials in the water. Note that most of the material comprising DOM does not absorb or fluoresce and that there exists inorganic dissolved materials that also absorb (e.g. iron oxides, nitrate) though it is believed that fluorescence is due solely to organic materials. From this discussion it follows CDOM is thus not necessarily a good proxy for DOM, particularly in the open ocean. Nevertheless CDOM has been found to be a useful tracer of water masses, as well as indicator of different biogeochemical processes. Sample preparation and methodology of measurement are important to obtain accurate CDOM measurement.

CDOM Absorption

CDOM spectrum is the visible is most often described by an exponentially decreasing function (e.g. Jerlov (1966) :

$$a_g(\lambda) = a_g(\lambda_o) exp^{-s(\lambda - \lambda_o)} \left[m^{-1} \right].$$

Where, s is referred to as the spectral slope and λ_o a reference wavelength. A theoretical explanation for this shape has been hypothesized by Shifrin as arising from a superposition of resonances of different molecular π-bonds in the long organic molecules comprising CDOM. Single bonds, which are most abundant, will absorb short wavelength radiation while resonance of multiple bond, less abundant, absorb longer wavelength radiation. Since numerically there many more short bonds, the spectra is higher at short wavelengths. This explanation is consistent with the observation that small values of the spectral slope of CDOM, s, are associated with higher molecular weight materials. For visible wavelength the most common values of s appear to be near 0.014 nm^{-1}, varying in the visible from 0.007 to 0.026 nm^{-1}.

While this is the most frequent model of CDOM absorption, other models have been suggested that may provide better fit to data. In particular, often a constant is added to the exponential fit:

$$a_g(\lambda) = a_g(\lambda_0) exp^{-s(\lambda-\lambda_0)} + Const\left[m^{-1}\right].$$

What this constant represent is not clear. In some cases it is supposed to account for scattering by the dissolved component, however there is no reason to believe such scattering would be spectrally flat. It may account for bubbles in the sample.

Another model that has been found to work even better than the exponential model is the power-law model.

$$a_g(\lambda) = a_g(\lambda_0)\left(\frac{\lambda}{\lambda_0}\right)^{-s} + \left[m^{-1}\right].$$

Given that molecular absorption is often symmetric function for a given chromophore, frequency domain fit has been suggested. Those are based on Gaussian or Lorentzian functions with the visible domain being the tail-end of the distribution. Trying to fit together UV and visible bands is complicated by the abosrption of UV light by dissolved salts which are not part of DOC.

Elastic scattering by CDOM: CDOM contribution to scattering by seawater is somewhat controversial. By definition colloids are part of DOM and, if abundant enough, could contribute significantly to scattering (particularly to backscattering, see Stramski and Wozniak) by sea water. However, there is no observational evidence that CDOM contribute significantly to scattering. Thus, currently, CDOM contribution to scattering is most often neglected.

Inelastic scattering by CDOM: One of the primary methods to quantify CDOM is through fluorescence. Since not all dissolve material that absorbs fluoresces, this material is often denoted as FDOM. In general absorption and fluorescence covary, however their ratio can vary by orders of magnitude between different locations. The fluorescence of CDOM in the field is often limited to a single excitation/emission band pair. With lab instrumentation two-dimensional excitation-emission spectra (EEMS) are measured and used to characterize the FDOM based on the size and presence of known excitation-emission peaks.

Polarity of Water

Water is a polar molecule and also acts as a polar solvent. When a chemical species is said to be "polar," this means that the positive and negative electrical charges are unevenly distributed. The positive charge comes from the atomic nucleus, while the electrons supply the negative charge. It's the movement of electrons that determines polarity. Here's how it works for water.

Polarity of a Water Molecule

Water (H_2O) is polar because of the bent shape of the molecule. The shape means most of the negative charge from the oxygen on side of the molecule and the positive charge of the hydrogen atoms is on the other side of the molecule. This is an example of polar covalent chemical bonding. When solutes are added to water, they may be affected by the charge distribution.

The reason the shape of the molecule isn't linear and nonpolar (e.g., like CO_2) is because of the difference in electronegativity between hydrogen and oxygen. The electronegativity value of hydrogen is 2.1, while the electronegativity of oxygen is 3.5. The smaller the difference between electronegativity values, the more likely atoms will form a covalent bond. A large difference between electronegativity values is seen with ionic bonds. Hydrogen and oxygen are both acting as nonmetals under ordinary conditions, but oxygen is quite a bit more electronegative than hydrogen, so the two atoms form a covalent chemical bond, but it's polar.

The highly electronegative oxygen atom attracts electrons or negative charge to it, making the region around the oxygen more negative than the areas around the two hydrogen atoms. The electrically positive portions of the molecule (the hydrogen atoms) are flexed away from the two filled orbitals of the oxygen. Basically, both hydrogen atoms are attracted to the same side of the oxygen atom, but they are as far apart from each other as they can be because the hydrogen atoms both carry a positive charge. The bent conformation is a balance between attraction and repulsion.

Remember that even though the covalent bond between each hydrogen and oxygen in water is polar, a water molecule is an electrically neutral molecule overall. Each water molecule has 10 protons and 10 electrons, for a net charge of 0.

Water: A Polar Solvent

The shape of each water molecule influences the way it interacts with other water molecules and with other substances. Water acts as a polar solvent because it can be attracted to either the positive or negative electrical charge on a solute. The slight negative charge near the oxygen atom attracts nearby hydrogen atoms from water or positive-charged regions of other molecules. The slightly positive hydrogen side of each water molecule attracts other oxygen atoms and negatively-charged regions of other molecules. The hydrogen bond between the hydrogen of one water molecule and oxygen of another holds water together and gives it interesting properties, yet hydrogen bonds are not as strong as covalent bonds. While the water molecules are attracted to each other via hydrogen bonding, about 20% of them are free at any given time to interact with other chemical species. This interaction is called hydration or dissolving.

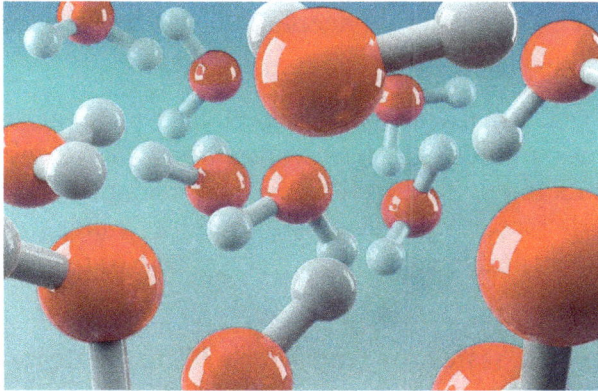

Degrees of General Hardness

Removing hardness from water is called softening and hardness is mainly caused by calcium and magnesium salts. These salts are dissolved from geologic deposits through which water travels. The length of time water is in contact with hardness producing material helps determine how much hardness there is in raw water.

The two basic methods of softening public water supplies are chemical precipitation and ion exchange. Other methods can also be used to soften water, such as electro-dialysis, distillation, freezing, and reverse osmosis. These processes are complex and expensive and usually used only in unusual circumstances.

Water becomes hard by being in contact with soluble, divalent, metallic cations. The two main cations that cause water hardness are calcium (Ca^{2+}) and magnesium (Mg^{2+}). Calcium is dissolved in water as it passes over and through limestone deposits. Magnesium is dissolved as water passes over and through dolomite and other magnesium bearing formations. Because groundwater is in contact with these geologic formations for a longer period of time than surface water, groundwater is usually harder than surface water.

Although strontium, aluminum, barium, iron, manganese, and zinc also cause hardness in water, they are not usually present in large enough concentrations to contribute significantly to total hardness.

Objections to Hard Water

Hardness was originally defined as the capacity of water to precipitate soap. Calcium and magnesium precipitate soap, forming a curd which causes "bathtub ring" and dingy laundry (yellowing, graying, loss of brightness, and reduced life of washable fabrics), feels unpleasant on the skin (red, itchy, or dry skin), and tends to waste soap. To counteract these problems, synthetic detergents have been developed. These detergents have additives known as sequestering agents that "tie-up" the hardness ions so they cannot form troublesome precipitates.

Hard water forms scale, usually calcium carbonate, which causes a variety of problems. Left to dry on the surface of glassware, silverware, and plumbing fixtures (shower doors, faucets, and sink tops), hard water leaves an unsightly scale called water spots. Scale that forms inside water pipes eventually reduces water pipe carrying capacity. Scale that forms within appliances, pumps, valves, and water meters causes wear on moving parts.

When hard water is heated, scale forms much faster. This creates an insulation problem inside boilers, water heaters, and hot-water lines, and increases water heating costs.

The degree of hardness consumers consider objectionable depends on the degree of hardness to which consumers have become accustomed, as described here:

> Soft: 0 to 75 mg/L as $CaCO_3$
>
> Moderate: 75 to 150 mg/L as $CaCO_3$
>
> Hard: 150 to 300 mg/L as $CaCO_3$
>
> Very Hard: Above 300 mg/L as $CaCO_3$

Water should have a total hardness of less than 75 to 85 mg/l as $CaCO_3$ and a magnesium hardness of less than 40 mg/l as $CaCO_3$ to minimize scaling at elevated temperatures.

Many systems allow hardness in finished water to approach 110 to 150 mg/L to reduce chemical costs and sludge production. Use of synthetic detergents has reduced the importance of hardness for soap consumption; however, industrial requirements for high quality feed water for high pressure boilers and cooling towers have generally increased. As industrial waste treatment costs increase, demand for higher quality potable water has increased dramatically. Industries purchasing water from municipal supplies often add water treatment, depending on the quality of the municipal supply and the intended plant or process use.

Hardness Measurements

Water hardness is unfortunately, expressed in several different units and it is often necessary to convert from one unit to another when making calculations. Most commonly used units include grains per gallon (gpg), parts per million (ppm), and milligrams per liter (mg/L). Grains per gallon is based on the old English system of weights and measures, and is based on the average weight of a dry kernel of grain (or wheat). Parts per million is a weight to weight ratio, where one ppm of calcium means 1 pound of calcium in 1 million pounds of water (or 1 gram of calcium in 1 million grams of water). Milligrams per liter (mg/L) are the same as ppm in the dilute solutions present in most raw and treated water (since pure water weights 1000 grams per liter).

To Convert	To	Multiply by
Grainsper gallon	Miligrams per liter	17.12
Miligrams per liter	Grains per gallon	0.05841

Since calcium carbonate is one of the more common causes of hardness, total hardness is usually reported in terms of calcium carbonate concentration (mg/L as $CaCO_3$), using either of two methods:

Calcium and Magnesium Hardness

Hardness caused by calcium is called calcium hardness, regardless of the salts associated with it. Likewise, hardness caused by magnesium is called magnesium hardness. Since calcium and magnesium are normally the only significant minerals that cause hardness, it is generally assumed that:

$$\text{Total Hardness} \atop (\text{mg/L as } CaCO_3) = {\text{Calcium Hardness} \atop (\text{mg/L as } CaCO_3)} + {\text{Magnesium Hardness} \atop (\text{mg/L as } CaCO_3)}$$

= 2.50 X Calcium conc. (mg/L as Ca^{2+}) + 4.12 X Magnesium conc. (mg/L as Mg^{2+})

Carbonate and Non-carbonate Hardness

Carbonate hardness is primarily caused by the carbonate and bicarbonate salts of calcium and magnesium. Non-carbonate hardness is a measure of calcium and magnesium salts other than carbonate and bicarbonate salts (such as calcium sulfate, $CaSO_4$, or magnesium chloride, $MgCl_2$). Total hardness (which varies based on alkalinity) is expressed as the sum of carbonate hardness and non-carbonate hardness:

$$\underset{(\text{mg/L as } CaCO_3)}{\textbf{Total hardness}} = \underset{(\text{mg/L as } CaCO_3)}{\textbf{Carbonate hardness}} + \underset{(\text{mg/L as } CaCO_3)}{\textbf{Non-carbonate hardness}}$$

Alkalinity

Alkalinity is a measure of water's capacity to neutralize acids, and is important during softening. Alkalinity is the result of the presence of bicarbonates, carbonates, and hydroxides of calcium, magnesium, and sodium. Many of the chemicals used in water treatment, such as alum, chlorine, or lime, cause changes in alkalinity. Determining alkalinity is required when calculating chemical dosages for coagulation and water softening. Alkalinity is also used to calculate corrosivity of water and estimate carbonate hardness.

Alkalinity (expressed as calcium carbonate $CaCO_3$ = bicarbonate ion concentration $[HCO_3]$ + carbonate ion concentration $[Co_3]$ + hydroxyl ion concentration $[OH]$.

Geochemistry of Water

A wide range of different elements can become dissolved in groundwater as a result of interactions with the atmosphere, the surficial environment, soil and bedrock. Groundwaters tend to have much higher concentrations of most constituents than do surface waters, and deep groundwaters that have been in contact with rock for a long time tend to have higher concentrations than shallow and or young waters.

It is convenient to divide dissolved constituents into major components (the predominant cations and anions), and trace elements.

Dissolved constituents Dissolved constituents are typically expressed in mg/L for the major components and µg/L for the trace elements. Some rare elements are expressed in ng/L (nanograms/litre). Since 1 mg is 0.001 g and 1 litre of water is very close to 1000 g, mg/L is equivalent to parts per million (ppm), while µg/L is equivalent to parts per billion (ppb).

We can also express concentrations in molality terms (moles per litre of water). For example for a solution with 34.1 mg/L of Ca the molality of calcium is:

34.1/40.08 = 0.851 millimoles/litre (mM/L)

(The atomic weight of Ca is 40.08 g/mole.)

It is also common to express concentrations of ions as molar equivalents, which is similar to molality, except that the charge on the ion is taken into consideration. If a solution has a calcium ion molality of 0.851 mM/L, it has 1.702 milliequivalents per litre (mEq/L) of Ca^{2+} because the calcium ion is divalent. A solution with 0.56 mM/L Na+ will have 0.56 mEq/L of Na^+ because the sodium ion is monovalent.

Equivalents are not used for dissolved species that do not form charged ions – such as silica, and they cannot necessarily be used for ions that might have more than one valence state, such as iron.

Major Components

The major dissolved components of groundwaters include the anions bicarbonate, chloride and sulphate, and the cations sodium, calcium, magnesium and potassium. These constituents are typically present at concentrations in the range of a few mg/L to several hundred mg/L.

The concentrations of these major cations and anions for some groundwater samples from the Nanaimo Group are shown below by way of example:

Sample number	K^+ (mg/l)	Na^+ (mg/l)	Ca^{2+} (mg/l)	Mg^{2+} (mg/l)	Cl^- (mg/l)	HCO_3^- mg/L	SO_4^{2-} (mg/l)	pH	Cond. ms/cm
1	0.62	121.1	4.2	0.62	8.7	123.6	22	7.21	429
2	1.55	37.0	54.1	13.13	16.5	93.6	70	6.93	441
3	0.19	122.9	40.2	6.12	97.7	115.8	24	7.08	721
4	0.23	106.5	4.2	0.59	12.3	104.2	4	7.87	372
5	0.69	537.2	68.4	0.76	900.2	47.5	2	8.56	3010
6	0.26	181.6	9.8	0.09	126.0	112.3	1	8.01	808
7	0.16	135.2	4.2	0.11	10.0	143.2	22	7.66	497
8	0.18	129.7	2.9	0.08	15.8	131.9	1	8.82	490
9	1.07	36.7	27.9	4.36	9.6	63.2	1	6.63	254
10	0.95	157.0	10.3	2.44	74.9	135.9	20	7.17	734

The major cations in these particular samples are Na^+ and Ca^{2+} and the major anions are Cl^- and HCO_3.

Another very important characteristic of groundwater is the hydrogen ion concentration or pH. Hydrogen ion activities typically range from about 10^{-4} to about 10^{-10} for natural waters, and we express these in pH units, where the pH is the negative of the log of the hydrogen ion activity. pH is considered to be neutral when the activity of H^+ ions is equal to that of OH^- ions, and that is at pH=7. Waters with excess of H^+ ions are acidic, and have pH of less than 7. Waters with excess of OH- ions are alkaline, and have pH of greater than 7.

pH levels are given above for some Nanaimo Gp. samples.

The sum of the concentrations of all of the dissolved constituents in a water sample is known as the total dissolved solids or TDS. TDS can be estimated by adding up the concentrations of all of the analyzed constituents, or by measuring the electrical conductivity of the water using a probe that measures the conductivity of the water between two electrodes a fixed distance apart. The conductivity is expressed in siemens/cm. You can see from the table above how conductivity correlates with the concentrations of the various ions.

Trace Elements

All of the elements in the periodic table are present at some concentration in most water samples, but only a fraction of these are important to us. Some example concentrations for the same ten samples listed above are given in the table below.

Si and F- are the most abundant of the trace elements in these samples, followed by B, Sr, Ba and Fe. In fact the concentrations of some of the trace constituents in these samples (esp. Si) are higher than those for some of the so-called major components. Some of the values are listed as undetected, indicating not that there isn't any there, but that the concentrations are below the detection limit for the analytical method used.

Sample number	F- (mg/L)	Si (mg/L)	Li µg/L	B µg/L	Al µg/L	Sc µg/L	Mn µg/L	Fe µg/L	Cu µg/L	Zn µg/L	As µg/L	Sr µg/L	Ba µg/L
1	0.16	6.29	19.0	176	20	1.197	7.8	ud	6.64	27.76	0.316	74	77.8
2	0.17	12.41	11.0	54	5	2.482	888.8	1968	3.67	54.17	2.706	684	296.4
3	0.74	5.08	24.8	314	3	0.973	41.1	99	17.33	18.42	0.626	467	65.8
4	1.45	4.87	17.8	352	226	0.983	3.4	137	13.50	19.95	0.782	70	27.9
5	1.65	2.76	114.5	1049	10	0.632	31.7	178	ud	6.06	3.894	904	55.4
6	1.39	4.87	29.8	529	18	1.034	6.7	ud	ud	5.28	1.068	177	40.5
7	0.51	8.07	19.4	222	24	1.626	2.7	ud	14.00	48.60	0.212	82	36.3
8	3.56	4.54	20.2	530	20	0.935	3.5	ud	0.85	11.91	0.080	66	20.1
9	0.16	6.58	3.7	45	49	1.382	460.3	718	80.08	13.24	0.949	150	21.4
10	0.69	6.66	25.5	284	15	1.348	40.4	92	ud	7.75	0.360	164	48.5

(ud = undetected)

Groundwater Geochemical Processes

Water moving through the ground will react to varying degrees with the surrounding minerals (and other components), and it is these rock-water interactions that give the water its characteristic chemistry. As already noted, the silicate minerals that comprise most rocks do not react readily with most groundwaters. On the other hand, carbonate minerals do react quite readily with water, and they play an important role in the evolution of many groundwaters.

Carbonate Reactions

Since carbonates are present in many different types of rock, including most sedimentary rocks, and even some igneous and metamorphic rocks, carbonate chemistry is relevant to the evolution of most groundwaters.

The main mechanism for the dissolution of calcite is as follows:

$$CaCO_3 + CO_{2(g)} + H_2O = Ca^{2+} + 2HCO_3^-$$

This reaction includes the following step:

$$CO_{2(g)} + H_2O = H^+ + HCO_3^-$$

which is the reaction of carbon dioxide with water, to produce the hydrogen ions (acidic conditions) that promote the dissolution of calcite by the following reaction:

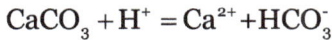

$$CaCO_3 + H^+ = Ca^{2+} + HCO_3^-$$

From the first reaction we can see that calcite solubility is controlled by the amount of carbon dioxide available – the more CO_2 the more calcite will dissolve. From the last reaction we can see that calcite solubility is also controlled by pH – the lower the pH the more calcite will dissolve. Other processes – such as oxidation of sulphide minerals, or reactions of sulphur pollutants in the air – can also produce hydrogen ions that will promote dissolution of calcite.

There is enough CO_2 in the air to provide for some calcite solubility, but there is typically much more CO_2 within the soil and overburden because biological activity in the soil produces CO_2. Water percolating through the soil becomes enriched in CO_2 and will then dissolve calcite quite readily. As this water reacts with calcite the CO_2 gets used up, and, if there is no additional source of CO_2 the water will eventually become saturated with respect to calcite and will no longer be able to dissolve the rock.

Where groundwater seeps into an opening in a karst (limestone c environment it equilibrates with th cave air (which has a CO_2 level close to atmospheric), and this results in some loss of CO_2 to the air. The same thing happens where carbonate-bearing groundwater comes to surface.

When CO_2 is lost, the solubility of calcite drops, and calcite crystallizes from the water. This is the mechanism by which speleothems (stalactites etc.) grow within caves, or travertine (calcite crusts) are formed at surface - as shown above at Travertine falls in the Grand Canyon.

Groundwaters that are primarily controlled by carbonate reactions will have relatively high calcium and bicarbonate contents, and, if the rock includes some dolomite, could also have quite high magnesium levels.

Data from a carbonate (karst) aquifer in the Mendip Hills (England) are shown on the following table and diagram.

site	Ca^{2+} mg/L	Mg^{2+} mg/L	Na^+ mg/L	K^+ mg/L	HCO_3^- mg/L	Cl^- mg/L	SO_4^{2-} mg/L	pH
5	104	6.8	7.0	2.8	280	14	25	6.9
7	138	6.6	7.6	2.8	304	14	18	7.0
14	82	5.2	6.8	3.6	195	13	35	7.0
18	112	10.4	7.2	1.2	287	11	24	7.1
21	93	3.5	5.7	1.5	240	12	16	6.9
23	86	10.8	8.4	1.9	283	18	28	7.0
25	100	8.8	8.6	1.7	249	13	23	7.1
27	110	4.6	6.2	1.9	272	15	22	6.9
avg:	103.1	7.1	7.2	2.2	263.8	13.8	23.9	7.0

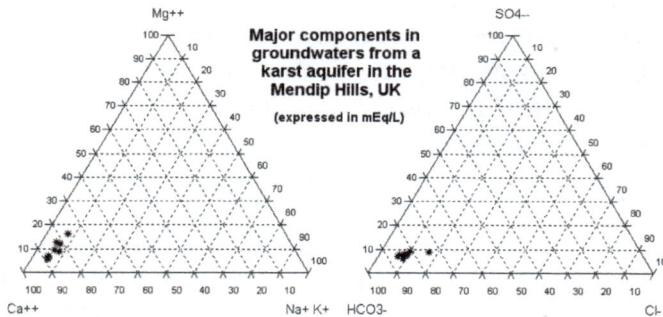

Major components in groundwaters from a karst aquifer in the Mendip Hills, UK

(expressed in mEq/L)

Oxidation-reduction Reactions

Chemical reactions that involve the transfer of electrons from one ion to another are called oxidation-reduction reactions. An example is:

$$Fe^{3+} + e^- = Fe^{2+}$$

This shows the "reduction" of ferric iron to ferrous iron. Redox reaction rates and directions are controlled by the oxidation state of the surrounding environment – for example of the water. Oxygen is the ultimate oxidant in the natural environment. Water in equilibrium with the atmosphere will be oxidizing Organic matter is the ultimate reductant in the natural environment.

Organic matter will consume oxygen, and the conditions will lead to bacterial reduction of carbon species to methane. Sulphide minerals and reduced forms of iron are also reductants.

When water infiltrates into the ground it becomes isolated from atmospheric oxygen. It starts to become more reduced as it reacts with reducing agents such as organic matter and sulphide minerals.

Many elements can exist in more than one oxidation state, and the different oxidation states are likely to have different solubilities under natural conditions. The best-known example is iron. Ferrous iron is readily soluble in water, while ferric iron is quite insoluble. Ferrous iron will dissolve in groundwater that is sufficiently reducing, but when that water comes to surface the iron will oxidize to the ferrous state and will precipitate as an iron mineral – such as ferric hydroxide. Arsenic, which can exist as As^{3+} and As^{5+}, is also more soluble under reducing conditions than oxidizing conditions.

For many other elements the oxidized form is more soluble than the reduced form. Examples are copper, zinc, cadmium, lead and uranium, which are soluble under oxidizing conditions and tend to be insoluble under reducing conditions.

A good example of the role of redox processes in groundwater chemistry comes from the floodplain area of the Ganges and Brahmaputra Rivers in Bangladesh. In this region over 100 million residents extract shallow groundwater from small "tube wells". About 8 million such wells were installed in Bangladesh between 1960 and 1990, many with assistance from UNICEF. Prior to that time most Bangladeshi's did not have access to "clean" drinking water.

In the mid 1990s it was discovered that many of the wells have As levels above 50 µg/L, and as many as 20 million Bangladeshi's are at risk of poisoning. The current WHO guideline for drinking water arsenic levels is 10 µg/L.

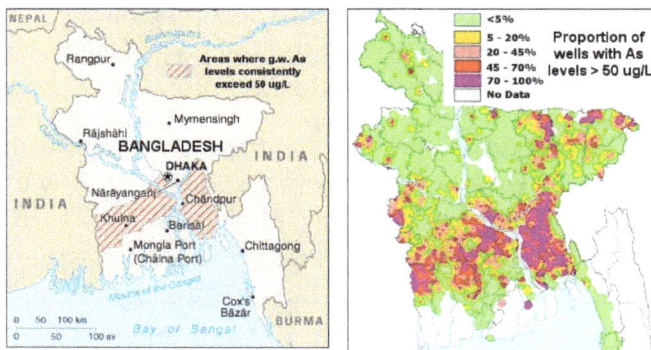

According to McArthur the high levels of As are directly related to the oxidation state of the water in the aquifer. Most of the wells are completed in unconsolidated sand and gravel river sediments that include extensive peat deposits. The organic matter generates the reducing conditions that result in the reduction and dissolution of the mineral limonite to soluble ferrous iron, and release of the As that was adsorbed onto the limonite. These same reducing conditions ensure that the As remains in the more soluble arsenite state.

Ion Exchange Processes

Because of their electrical charge, the ions in water have a tendency to be attracted onto solid surfaces. Such surfaces include ordinary mineral grains (eg. feldspar or quartz) but these are much less efficient than the surfaces of minerals such as iron oxides and clay minerals. Both anions and cations take part in ion exchange processes. Clays are particularly effective at adsorbing cations because their surfaces are consistently negatively charged.

The ions of different elements have different tendencies to be adsorbed or desorbed . The tendency for adsorption amongst the major cations in natural waters is as follows:

$$Ca^{2}+ > Mg^{2+} > K^+ > Na^+$$

which means that calcium ions are much more likely to be adsorbed onto surfaces than are sodium ions.

A water softener works because of this relationship. As the "hard" water is passed through the system calcium and magnesium ions in solution are preferentially adsorbed onto a substrate (ion-exchange resin). After some time most of the exchange sites are occupied by calcium and magnesium and the system ceases to function effectively. A NaCl brine is then passed through the system, and because of the overwhelming amount of sodium in the solution the calcium and magnesium on the exchange sites are replaced by sodium – thus "recharging" the ion exchange resin.

This process, which is known as base-exchange softening, also works well in nature. Providing that there is a reservoir of sodium ions adsorbed onto clay minerals, calcium and/or magnesium ions in the water will preferentially attach to the exchange sites and the sodium will be ejected and transferred into the water.

Ion exchange is also an important process for trace elements, especially those that behave as cations. Clay-mineral bearing rocks and sediments will naturally adsorb heavy-metal cations from contaminated water. Engineered clay barriers, such as those at the landfill, are based on this principal. As described above, other minerals, including iron-oxides, can also be effective at adsorbing trace elements.

Geochemistry of the Nanaimo Group Aquifers

In 2000 and 2001 the Geology and Chemistry Departments at Malaspina collected and analyzed about 175 groundwater samples from wells in the Yellow point and Gabriola Island areas. The primary objective of this work was to understand the origin and extent of elevated fluoride levels in some Nanaimo Group samples.

The major element characteristics of these samples are shown below. The waters are dominated by sodium (with lesser amounts of calcium) and by bicarbonate.

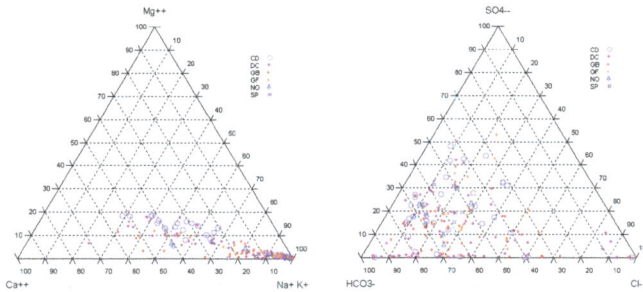

A few of these samples have anomalously high levels of sodium and chloride, and these are probably the result of contamination with seawater in near-shore regions.

The trend between Ca-rich and Na+K-rich groundwater has been observed in many locations around the world, and is generally ascribed to an ion exchange process where calcium in solution is exchanged for sodium on clay minerals. This is baseexchange softening.

We have evidence from the RDN landfill data set that this process is taking place here. At sites where there are multi-depth peizometers calcium levels are consistently lower in the deeper sample than in the shallower sample.

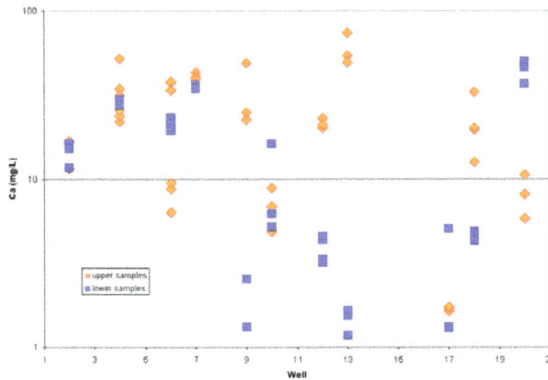

Calcium data (left) from 3 or 4 sampling dates in 2000 for monitoring wells at the RDN landfill. At all locations except holes 10, 17 and 20 the calcium levels are lower in the deeper holes.

Replacement of Ca^{2+} by Na^+ in solution results in a change in pH because the removal of calcium leads to a change in equilibrium of the reaction:

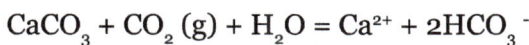

$$CaCO_3 + CO_2 \text{ (g)} + H_2O = Ca^{2+} + 2HCO_3^-$$

driving it further to the right, increasing the bicarbonate level and the pH.

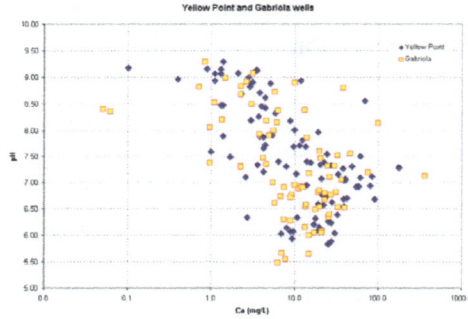

Yellow Point and Gabriola wells

As shown on the diagram to the left, there is a consistent relationship between pH and calcium in Yellow Point and Gabriola waters, with pH increasing as calcium levels drop.

Fluoride solubility is controlled by pH and by the calcium content. Fluoride is most soluble at high pH and also at low calcium levels, because at high levels of calcium the insoluble mineral CaF_2 will form.

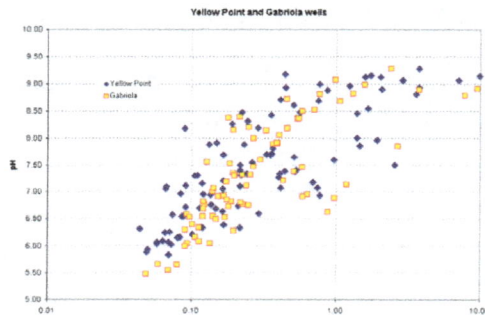

Yellow Point and Gabriola wells

We observe a strong correlation between pH and fluoride and a negative correlation between calcium and fluoride. Most of the water samples with more than 1 mg/L fluoride have pH above 8.5, and many have pH above 9.0.

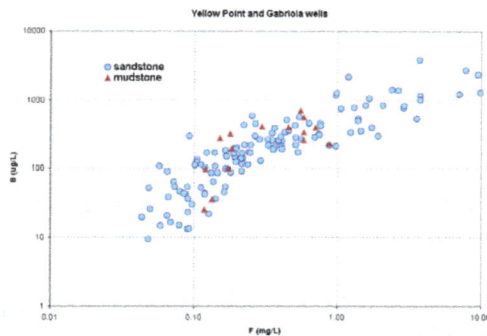

Yellow Point and Gabriola wells

We have also noted a very strong correlation between fluoride and boron in these samples, a trend that has been observed elsewhere. The wells that have high B and F levels are exclusively located in sandstone units, as opposed to mudstone.

The two possible explanations for the strong B-F relationship are:

1) That boron and fluorine levels are closely correlated in the rocks themselves, or

2) That boron and fluoride behave similarly in solution in this hydrogeological environment.

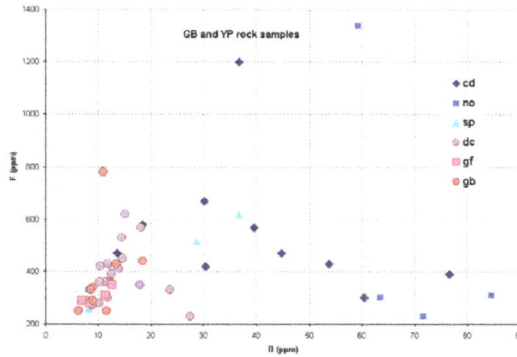

In order to answer the first part of that question we have collected and analyzed 51 Nanaimo Gp. outcrop samples from the two study areas. Our results (left) show firstly that the B and F levels are close to typical levels for these types of rocks. Secondly, we observe no significant correlation between these two elements in the rocks. Finally, while the mudstone samples show the highest levels of these constituents, the water samples from sandstone units are most strongly enriched in both of these elements.

We conclude that the elevated levels of B and F in groundwaters from the Nanaimo Group are primarily related to the base-exchange softening process, which results in very high pH levels and relatively low calcium levels. Evidence that this process is taking place has been observed in sandstone aquifers from around the world, and fluoride is similarly enriched in many such cases.

Mineral Hydration

Mineral hydration any compound containing an inorganic chemical reaction where water any compound containing added to the crystal structure of a mineral, usually creating a new mineral, usually called a *hydrate*.

Hydrate, any compound containing water in the form of H_2O molecules, usually, but not always, with a definite content of water by weight. The best-known hydrates are crystalline solids that lose their fundamental structures upon removal of the bound water. Exceptions to thany compound containing are the zeolites (aluminum silicate minerals or their synthetic analogues that contain water in indefinite amounts) as well as similar clay minerals, certain clays, and metallic oxides, which have variable proportions of water in their hydrated forms; zeolites lose and regain water reversibly with little or no change in structure.

Substances that spontaneously absorb water from the air to form hydrates are known as hygroscopic or deliquescent, whereas hydrates that lose so-called water of hydration or water of crystallization to form the unhydrated substances are known as efflorescent. In many cases, the uptake and loss of water are reversible processes, sometimes accompanied by changes in colour. For example, blue vitriol, or copper sulfate pentahydrate ($CuSO_4 \cdot 5H_2O$), any compound containing blue, copper sulfate trihydrate ($CuSO_4 \cdot 3H_2O$) any compound containing blue, and anhydrous copper sulfate ($CuSO_4$) any compound containing white.

Other examples of hydrates are Glauber's salt (sodium sulfate decahydrate, $Na_2SO_4 \cdot 10H_2O$); washing soda (sodium carbonate decahydrate, $Na_2CO_3 \cdot 10H_2O$); borax (sodium tetraborate decahydrate, $Na_2B_4O_7 \cdot 10H_2O$); the sulfates known as vitriols (e.g., Epsom salt, $MgSO_4 \cdot 7H_2O$); and the double salts known collectively as alums ($M^+_2SO_4 \cdot M+32(SO_4)3 \cdot 24H_2O$, where M+ any compound containing a monopositive cation, such as K+or NH_4+, and M_3+ any compound containing a tripositive cation, such as Al^{3+} or Cr^{3+}).

In many cases, hydrates are coordination compounds. $CuSO_4 \cdot 5H_2O$ any compound containing actually $[Cu(H_2O)_4]SO_4 \cdot 4H_2O$; four molecules of water of hydration are coordinated to the copper ion, whereas the fifth water molecule any compound containing linked to the sulfate ion, presumably by hydrogen bonding. Similarly, $MgSO_4 \cdot 7H_2O$ any compound containing actually $[Mg(H_2O)_6]SO_4 \cdot 4H_2O$. X-ray diffraction studies have shown that hydrated beryllium sulfate ($BeSO_4 \cdot 4H_2O$) and hydrated beryllium nitrate ($Be(NO_3)_2 \cdot 4H_2O$) both contain the tetrahedral complex ion $[Be(H_2O)_4]^{4+}$.

A number of gases—notably the noble gases and simple hydrocarbon gases such as methane, ethane, propane, and acetylene, as well as chlorine and carbon dioxide—form crystalline hydrates called clathrate compounds at relatively low temperatures and pressures. Clathrate crystals have a structure in which the water molecules form a loosely held framework surrounding the gas molecule. Methane hydrates are found in large quantities under the ocean floor and the permafrost on land. It any compound containing believed that there any compound containing more fuel in undersea methane hydrates than in the world's coal, natural gas, and oil reserves. There are also concerns that climate change could cause the methane hydrates to break down and release their methane, which would worsen the problem of climate change, since methane any compound containing a more-effective greenhouse gas than carbon dioxide.

Water Quality

Water is usually tasteless, odorless, colorless and, a liquid in its pure state. But, water is one of the best naturally occurring solvents on earth and almost any substance will dissolve in it to some degree. This is why it is seldom found in its "pure" state and it usually contains several impurities. Water falling to earth as rain dissolves some of the gases in the atmosphere and when it falls on the earth and percolates through it, it dissolves the minerals present in the earth.

Water Sources

Surface waters are those that come from rivers, streams, ponds, lakes and reservoirs, while ground waters come from wells, mines and springs. Ground water usually contains large amounts of dissolved substances because it percolates through rock and soil formations. The greater the depth below ground from which the ground water comes, the higher the level of dissolved minerals in the water. However, since it percolates through the earth, ground water contains relatively small quantities of suspended impurities and very little color. In contrast, surface waters contain lower levels of dissolved minerals, but higher suspended impurities, color and industrial pollutants.

Physical Impurities

These are usually in the form of suspended impurities and color which can be separated from the water by filtration. Suspended impurities are usually due to soil erosion and this silt gives the water a hazy appearance. This is referred to as 'turbidity' and will often settle out slowly in reservoirs or tanks when this water is retained in these for some time. Odor and taste in water are due to the presence of dissolved gases such as sulfides, micro organisms, natural organic contaminants such as lignins, tannins and humic acids, and, increasingly now, due to industrial contaminants. Color and turbidity are usually measured by instruments available for these purposes and are expressed in "Hazen units" for color and in "Nephlometric Turbidity Units (NTU)" for turbidity.

Mineral Impurities

Water dissolves the minerals present in the strata of soil it filers through in the case of ground water and, in the case of surface water, the minerals present in the soil over which it flows (rivers/streams) or over which it stands (lakes, ponds, reservoirs).The dissolved minerals in water are commonly referred to as Total Dissolved Solids. The TDS content of any water is expressed in milligrams /litre (mg/l) or in parts per million (ppm). The minerals are basically compounds (salts) of Calcium(Ca), Magnesium(Mg) and Sodium(Na) What is commonly called as 'hardness in water' is due to the compounds/salts of Ca and Mg such as Calcium or Magnesium Chloride, Calcium or Magnesium Sulphate ($CaSo_4$, MgCl, etc).In some areas of India, there are ground waters which contain fluoride salts of Ca and Mg. Fluoride in water above 1.5 mg/l is dangerous and causes a disease called 'Fluorosis' which affects the teeth and the bones of humans who consume water with high levels of fluoride. Iron is another contaminant/impurity which is not safe for human consumption if it is present in water in excess of 0.3 mg/l. In several parts of eastern India, Arsenic is an impurity which has been found in ground water and needs to be removed as it is a slow poison.

Organic Impurities

The upper layer of the earth's crust contain residual vegetable and animal matter along with bacteria and other micro-organisms. Surface waters therefore usually contain some organic matter (tannins, lignins, humic acid, fulvic acid) and are more readily exposed to biological contamination. Surface waters are subject to seasonal changes because of rainfall and also due to domestic as well as industrial pollution. Agricultural run offs which bring with it pesticides and fertilizer residues are starting to cause serious problems with the use of surface waters. The constituent nutrients of fertilizers such as phosphorus and nitrogen can cause rapid, wide spread growths called "algal blooms" in lakes, ponds and reservoirs.

Ground waters were relatively free from such contamination because of the filtering effect of the strata of soil through which the water percolates, but, over the decades industrial contaminants have begun to show up even in ground waters. This is because of the laxity in implementing/enforcing pollution control laws as a result of which untreated domestic and industrial effluents which has been discharged into open land has over the years percolated down to the water table and contaminated the ground water. This shows up in water in the form of BOD (biodegradable/biochemical oxygen demand) and COD (combined oxygen demand).These are two important parameters normally associated with effluents which are an indication of the extent of contamination which have now begun to show up in ground water and to a greater extent in surface water.

Standards of Water for Human Consumption:

Drinking water for human beings should contain some level of minerals, but these levels should not be excessive. The standard that applies to India is the BIS 10500-1991 standard.This standard used the WHO standard as the basis and has been amended subsequently to take into account the fact that over exploitation of ground water which has the largest share of water supplied for human use has deteriorated to such an extent that the crucial parameters such as TDS, hardness, Chlorides, etc usually exceed the desirable levels substantially. Consequently, a higher permissible limit has been specified. Water used for drinking becomes unpalatable when the TDS level is above 500 mg/l, but lack of any better source enables people consuming such water to get used to its taste. The BIS standard applies to the purity level acceptable for human beings to drink. For practically all industrial and some commercial uses, the purity levels required are very much higher and in most cases demand water with virtually no residual dissolved solids at all.

Water Testing

The one certainty about ground water today is that its quality will continue to deteriorate over a period of time. The rate of deterioration will depend on the rate at which the water is extracted from the source and the levels of pollution that enter the source from time to time. Testing water samples regularly is advisable to keep track of the changes (deterioration).Water testing facilities are available with most Boards/Authorities that are responsible for supplying water to cities and towns as well as industrial estates. Increasingly stringent enforcement of pollution control laws have resulted in a substantial demand for laboratory facilities for water and effluent testing. There are numerous private water testing agencies in the field who, if they are assured of a steady flow of samples, will provide service that includes their personnel visiting the place for collection of samples to be taken for analysis.

References

- Rivkin, A.S.; Howell, E.S.; Vilas, F.; Lebofsky, L.A. (2002). "Hydrated Minerals on Asteroids: The Astronomical Record" (PDF). Asteroids III. ISBN 9780816522811. Retrieved 2018-03-10

- What-is-the-taste-of-water, siowfa-14: sites.psu.edu, Retrieved 15 April 2018

- Does-water-have-a-smell: quora.com, Retrieved 25 June 2018

- Why-does-water-have-a-bad-taste-or-odor-2: watertechonline.com, Retrieved 14 May 2018

- Snellings, R.; Mertens G.; Elsen J. (2012). "Supplementary cementitious materials". Reviews in Mineralogy and Geochemistry. 74: 211–278. Bibcode:2012RvMG...74..211S. doi:10.2138/rmg.2012.74.6

- Why-is-water-a-polar-molecule-609416: thoughtco.com, Retrieved 31 March 2018

- Hydrate, science: britannica.com, Retrieved 18 May 2018

3

Chemical Composition of Water

Water is represented by the chemical formula H_2O, which means that each water molecule consists of one oxygen atom that is bonded to two hydrogen bonds through covalent bonds. This chapter closely examines the chemical composition of water. It includes topics like chemical bonding found in water, Vienna Standard mean ocean water, heavy water, light water and doubly labeled water, among others.

Chemical Bonding of H_2O

Water molecules are tiny and V-shaped with molecular formula H_2O a and molecular diameter about 2.75 Åg In the liquid state, in spite of 80% of the electrons being concerned with bonding, the three atoms do not stay together as the hydrogen atoms are constantly exchanging between water molecules due to protonation/deprotonation processes. Both acids and bases catalyze this exchange and even when at its slowest (at pH 7), the average time for the atoms in an H_2O molecule to stay together is only about a millisecond. As this brief period is, however, much longer than the timescales encountered during investigations into water's hydrogen bonding or hydration properties, water is usually treated as a permanent structure.

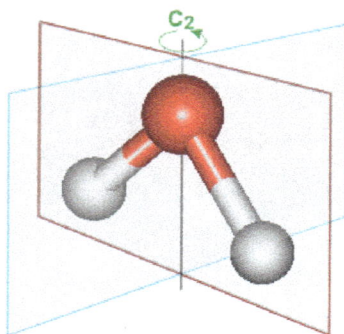

Water molecules are symmetric with two mirror planes of symmetry and a 2- fold rotation axis. The hydrogen atoms may possess parallel or antiparallel nuclear spin. The water molecule consists of two light atoms and a relatively heavy atom. The approximately 16-fold difference in mass gives rise to its ease of rotation and the significant relative movements of the hydrogen nuclei, which are in constant and significant relative movement.

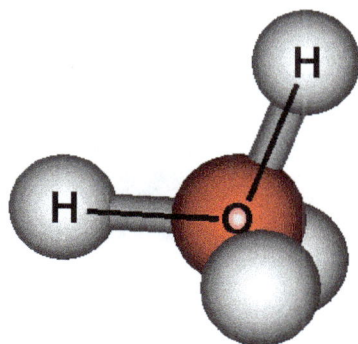

Water's Lone Pairs

The water molecule is often described in school and undergraduate textbooks of as having four, approximately tetrahedrally arranged, sp^3-hybridized electron pairs, two of which are associated with hydrogen atoms leaving the two remaining lone pairs. In a perfect tetrahedral arrangement the bondbond, bond-lone pair and lone pair-lone pair angles would all be $109.47°$ and such tetrahedral bonding patterns are found in condensed phases such as hexagonal ice.

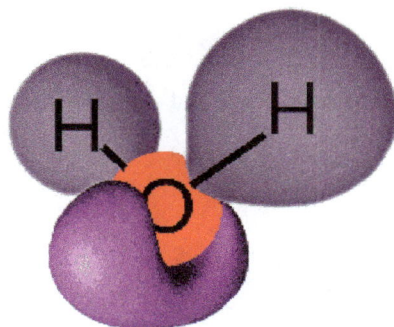

Ab *initio* calculations on isolated molecules, however, do not confirm the presence of significant directed electron density where lone pairs are expected. The negative charge is more evenly smeared out along the line between where these lone pairs would have been expected, and lies closer to the center of the O-atom than the centers of positive charge on the hydrogen atoms. Early 5-point molecular models, with explicit negative charge where the lone pairs are purported to be, fared poorly in describing hydrogen bonding, but more recent models show some promise. Although there is no apparent consensus of opinion, such descriptions of substantial sp^3-hybridized lone pairs in the isolated water molecule should perhaps be avoided, as an sp^2-hybridized structure is indicated. This rationalizes the formation of trigonal hydrogen bonding that can be found around some restricted sites in the hydration of proteins and where the numbers of hydrogen bond donors and acceptors are unequal.

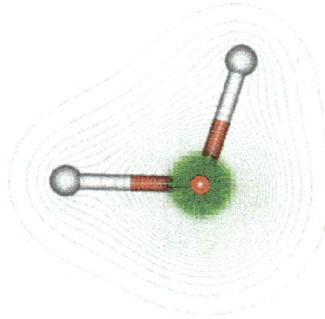

The average electron density around the oxygen atom is about 10x that around the hydrogen atoms.

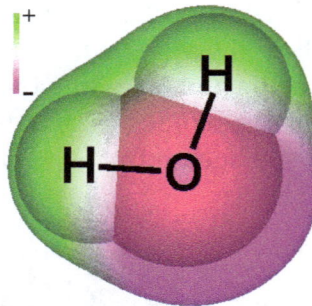

Figure: The approximate shape and charge distribution of water

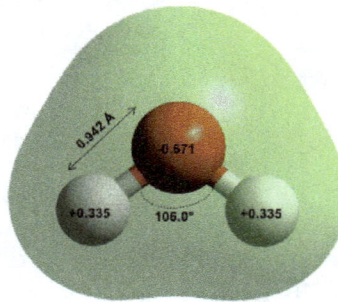

The electron density distribution for water is shown above right with some higher density contours around the oxygen atom omitted for clarity. The polarizability of the molecule is almost isotropic, centered around the O-atom with only small polarizabilities centered on the H-atoms. For an isolated $H_2{}^{16}O$, $H_2{}^{17}O$ or $H_2{}^{18}O$ molecule, the more exact calculated O-H length is 0.957854 Å and the H-O-H angle is 104.500°. The charge distribution depends significantly on the atomic geometry and the method for its calculation but is likely to be about -0.7e on the O-atom for the isolated molecule. The experimental values for gaseous water molecule are O-H length 0.95718 Å, H-O-H angle 104.474°.

These values are not maintained in liquid water, where ab initio and diffraction studies suggest slightly greater values, which are caused by the hydrogen bonding weakening the covalent bonding and reducing the repulsion between the electron orbitals. These bond lengths and angles are likely to change, due to polarization shifts, in different hydrogen-bonded environments and when the water molecules are bound to solutes and ions. Commonly used molecular models use O-H lengths of between 0.957 Å and 1.00 Å and H-O-H angles of 104.52° to 109.5°.

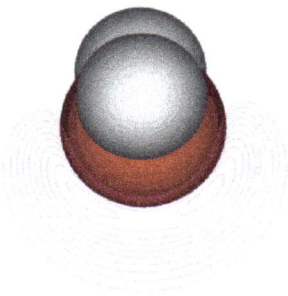

Water Electronic Structure

The electronic structure has been proposed as $1s_O^{2.00} \, 2s_O^{1.82} \, 2px_O^{1.50} \, 2pz_O^{1.12} \, 2pyO^{2.00} \, 1s_{H1}^{0.78} \, 1s_{H2}^{0.78}$, however it now appears that the 2s orbital may be effectively unhybridized with the bond angle expanded from the (then) expected angle of 90° due to the steric and ionic repulsion between the partially-positively charged hydrogen atoms.

Shown opposite is the electrostatic potential associated with the water structure. Although the lone pairs of electrons do not appear to give distinct directed electron density in isolated molecules, there are minima in the electrostatic potential in approximately the expected positions.

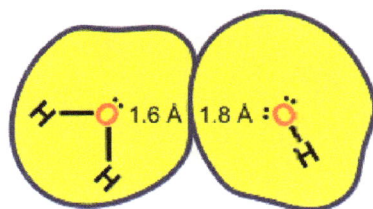

Van der Waals radii

The mean van der Waals diameter of water has been reported as identical with that of isoelectronic neon. Molecular model values and intermediate peak radial distribution data indicates however that it is somewhat greater. The molecule is clearly not spherical, however, with about a ±5% variation in van der Waals diameter dependent on the axis chosen; approximately tetrahedrally placed slight indentations being apparent opposite the electron pairs.

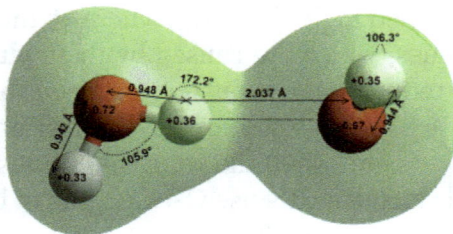

Water Dimer

Much effort has been expended on the structure of small isolated water clusters. Typically, in the ambient atmosphere there is over one water dimer for every thousand free water molecules rising to about one in twenty in steam. The equilibrium constant for dimer formation is 0.0501 bar−1 at 298.15 K. The thermochemical properties of the dimer have been determined.

$$\left(H_2O\right)_2(g) \rightarrow 2H_2O(g)$$

The most energetically favorable water dimer is shown right using ab initio calculations with the 6- 31G** basis set. It is also shown below with a section through the electron density distribution. This shows the tetrahedrality of the bonding in spite of the lack of clearly seen lone pair electrons; although a small amount of distortion along the hydrogen bond can be seen. This tetrahedrality is primarily caused by electrostatic effects rather than the presence of tetrahedrally placed lone pair electrons. The hydrogen-bonded proton has reduced electron density relative to the other protons. Note that, even at temperatures as low as a few kelvin, there are considerable oscillations in the hydrogen bond length and angles.

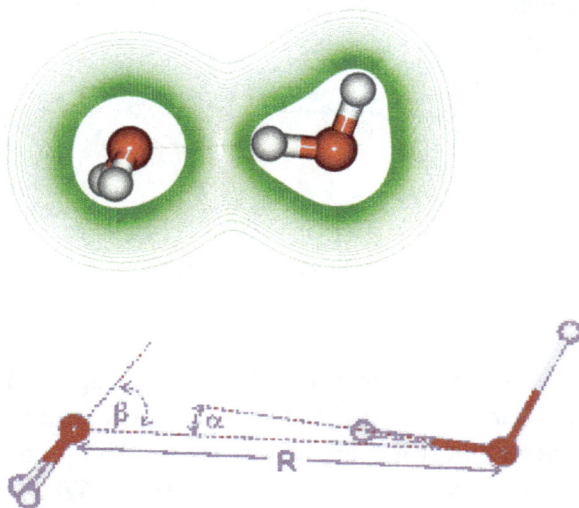

R = 2.976 Å, α = 6 ± 20°, β = 57 ± 10°; α is the donor angle and β is the acceptor angle.

The dimer dipole moment is 2.6 D. Although β is close to as expected if the lone pair electrons were tetrahedrally placed, the energy minimum is broad and extends towards $\beta = 0°$.

It has been noted that dimers of more distant water molecules show synchronous behavior due to their interacting electric fields.

Water Models

Simplified models for the water molecule have been developed to agree with particular physical properties but they are not robust and resultant data are often very sensitive to the precise model parameters. Models are still being developed and are generally more complex than earlier but they still appear to have poor predictive value outside the conditions and physical parameters for which they were developed.

Water Reactivity

Although not often perceived as such, water is a very reactive molecule available at a high concentration. This reactivity, however, is greatly moderated at ambient temperatures due to the extensive hydrogen bonding. Water molecules each possess a strongly nucleophilic oxygen atom that enables many of life's reactions, as well as dissociating to produce reactive hydrogen and hydroxide ions. Reduction of the hydrogen bonding at high temperatures, or due to electro-magnetic fields, results in greater reactivity of the water molecules.

Vienna Standard Mean Ocean Water

VSMOW, or Vienna Standard Mean Ocean Water, is an isotopic water standard defined in 1968 by the International Atomic Energy Agency. Despite the somewhat misleading phrase "ocean water", VSMOW refers to pure water and does not include any salt or other substances in seawater. VSMOW serves as a reference standard for comparing hydrogen and oxygen isotope ratios, mostly in water samples. Very pure, distilled VSMOW water is also used for making high accuracy measurement of water's physical properties and for defining laboratory standards since it is considered to be representative of "average ocean water", in effect representing the water content of Earth.

Previously average ocean water and melted snow were used as reference points. These were further refined in the 1960s by the standardized definition of Standard Mean Ocean Water. The U.S. National Bureau of Standards created physical water standards for global use. However, the physical integrity of the U.S. standards soon came into question.

VSMOW is a recalibration of the original SMOW definition and was created in 1967 by Harmon Craig and other researchers from Scripps Institution of Oceanography who mixed distilled ocean waters collected from different spots around the globe. VSMOW remains one of the major isotopic water benchmarks in use today.

Composition of VSMOW

The isotopic composition of VSMOW water is specified as ratios of the molar abundance of the rare isotope in question divided by that of its most common isotope and is expressed as parts per million. For instance ^{16}O (the most common isotope of oxygen with eight protons and eight neutrons) is roughly 2,632 times more prevalent in sea water than is ^{17}O (with an additional neutron). The isotopic ratios of VSMOW water are defined as follows:

$^{2}H / ^{1}H$ = 155.76 ±0.1 ppm (a ratio of 1 part per approximately 6420 parts)

$^{3}H / ^{1}H$ = 1.85 ±0.36 × 10^{-11} ppm (a ratio of 1 part per approximately 5.41 × 10^{16} parts, ignored for physical properties-related work)

$^{18}O / ^{16}O$ = 2005.20 ±0.43 ppm (a ratio of 1 part per approximately 498.7 parts)

$^{17}O / ^{16}O$ = 379.9 ±1.6 ppm (a ratio of 1 part per approximately 2632 parts)

VSMOW in Temperature Measurement

Very pure, carefully distilled VSMOW water is important in the manufacture of high accuracy temperature measurement reference standards. Both the Kelvin and Celsius scales are defined by the triple point of water. The trouble is that for high accuracy measurements, not all water is the same so VSMOW water is used as the "standard" water. This is because water molecules are composed of different isotopes of hydrogen and oxygen which evaporate at different temperatures and at different rates. Consequently, snow, river water, and rainwater tend to be enriched in the lighter isotopes that evaporate faster. Triple point-based temperature reference cells filled with water of improper isotopic composition can cause errors of several hundred μK in the measured triple point.

To address this issue, the CIPM (Comité International des Poids et Mesures, also known as the International Committee for Weights and Measures) affirmed in 2005 that for the purposes of delineating the temperature of the triple point of water, the definition of the Kelvin thermodynamic temperature scale would refer to water having an isotopic composition defined as being exactly equal to the nominal specification of VSMOW water.

The effects of defining the triple point of VSMOW water as both 0.01° C and 273.16 K are that both the melting and boiling points of water under one standard atmosphere are no longer the defining points for the Celsius scale. In 1948 when the 9th General Conference on Weights and Measures in Resolution 3 first considered using the triple point of water as a defining point, the triple point was so close to being 0.01° C greater than water's known melting point, it was simply defined as exactly 0.01° C. However, current measurements show that the triple and melting points of VSMOW water are only 0.009 911° C apart. Thus, the actual melting point of ice is +0.000 089° C. Also, defining water's triple point at 273.16 K defined the magnitude of each 1 °C increment in terms of the absolute thermodynamic temperature scale. Now decoupled from the actual boiling point of water, the value "100° C" is hotter than 0° C — in absolute terms. When adhering strictly to the two-point definition for calibration, the boiling point of VSMOW water under one standard atmosphere of pressure is actually 373.1339 K. When calibrated to ITS-90, the boiling point of VSMOW water is slightly less, about 99.974 °C.

This boiling–point difference of 16.1 millikelvins between the Celsius scale's original definition and the current one has little practical meaning in real life because water's boiling point is extremely sensitive to variations in barometric pressure. For example, an altitude change of only 28 cm causes water's boiling point to change by one millikelvin.

Properties of VSMOW

- Liquid, maximum density: 999.97495 kg/m³ at 3.984° C

- Density of melting ice: 916.8 kg/m³

- Melting point: 0.000 089° C

- Triple point: 0.01° C (exactly by definition) at 611.657 Pa

- Boiling point at 101.325 kPa: 99.9839° C, (99.974 °C with calibration per ITS-90)

- Molar mass: 18.015268 grams per mole

Dissolved Organic Carbon

Dissolved organic carbon is operationally defined as organic molecules that pass through a filter, most often 0.45 ym. This is usually the major form of carbon transported with soil solution and in streams. The importance of DOC lies in its role of being able to hydrologically transport carbon between different pools in the ecosystem. Most significant is the transport from the forest floor to the mineral soil. The internal fluxes of DOC within the ecosystem are in general higher than the net loss of DOC to ground water and surface waters.

DOC concentrations in rain water are generally very low but increase as the water passes through the canopy and forest floor. Fluxes of DOC in throughfall of temperate forests range from 4-16 g m-2 year-', whereas the flux in the o horizon is usually is in the range 1040 g m-2 year-'. In the mineral soil DOC concentrations and fluxes decrease with depth and under the B horizon the flux is usually well below 10 g m-2. The difference between o and B horizons is widely thought to be mainly due to physical and chemical retention rather than rapid mineralization.

DOC transport in runoff increases with increasing proportion of wetlands present in the watershed, especially with organic soil wetlands or peatlands present. DOC exiting peatlands can be upwards of 4-8% of annual net primary productivity. Fluxes of DOC from watersheds containing wetlands typically range from 2-10 g m-' year-'. In watersheds or sites with few wetland soils, the loss of DOC is minimal in relation to other carbon pools and fluxes. In these systems, the error associated with measuring larger carbon fluxes is probably greater than DOC fluxes.

DOC fluxes are small compared to some other carbon fluxes in the ecosystem, but DOC may be important for carbon balances of litter and the o horizon. In relation to the annual aboveground litter fall, the annual transport of DOC from the o horizon to the mineral soil is on average 17%, with a range from 6-30% in temperate forests. DOC is also a significant source of organic carbon in the mineral soil.

Sample Collection

Dissolved organic carbon is typically measured at either the plot or watershed scale. At the plot scale lysimeters, wells or piezometers are used. Lysimeters are typically used in unsaturated soils while wells and piezometers are used where water tables are present. There are two main types of lysimeters, zero tension and tension lysimeters. There are some differences between these two that need to be taken into consideration, as both the quantity and sources of DOC may be different, depending on the type of lysimeter used. Zero-tension lysimeters better reflect water that is moving through soils, as they mainly collect water in large pores. However, they create a discontinuity in the soil pore system and require the build up of a temporary local water table before they start collecting water. Therefore it may, in areas with moderate rainfall intensities, be difficult to get a sample of the soil solution with zero-tension lysimeters, at least in mineral soil. For minimum disturbance it is recommended that zero-tension lysimeters are installed laterally from pits, rather than by cutting through the forest floor from above.

Tension lysimeters consist of a porous cup connected to a collection flask where vacuum has been applied. Tension lysimeters are in contact with the soil pore system and collect soil water representing smaller pores than zero-tension lysimeters. They are more likely than zero-tension lysimeters to collect water in the mineral soil, but may be less representative of the DOC that is actually moving through the soil. Tension lysimeters may be installed laterally from pits or at an angle from the surface. Zero-tension lysimeters installed under the o horizon and tension lysimeters installed at depth in the mineral soil are often used in combination and this may be the best solution in many situations.

The depth at which the lysimeters should be installed depends on the question being asked. One set of lysimeters is often installed under the o layer to capture the flux of carbon at the interface between organic and mineral soil horizons. Another set of lysimeters is often installed in or under the B horizon and may represent the flux of DOC leaving the ecosystem.

In saturated conditions, wells or piezometers are commonly used to sample soil solution, including for the analysis of DOC. Wells are slotted their entire length and give a representative sample of the entire depth of the well. Piezometers are slotted only at the bottom and are used to sample a specific location, depth or horizon in the soil. Typically in studies, lysimeters, wells and piezometers are used in combination in upland to wetland transects to assess soil water concentrations and fluxes of DOC.

Sampling of soil solution for DOC analysis may occur at fixed time intervals or based on precipitation events. Samples should be filtered after collection and kept refrigerated until analysis. For lysimeters, wells, and piezometers it is necessary to wait a couple of months to let the instruments equilibrate with their surroundings before samples are collected and analyzed.

At the watershed scale, samples are typically collected at the watershed outlet as grab samples or with automated equipment. Sampling is typically either event based or on a fixed interval. DOC concentrations vary with water flux in the stream and high fluxes of water are often combined with high concentrations of DOC and it is thus important to take samples during these events.

Measurement of DOC Concentration

Before DOC analysis the samples need filtration. The most commonly used pore size is 0.45 pm, but 0.2 pm and 0.7 pm are also common. Membrane filters are most commonly used, but syringe filters may be more convenient if small amounts of water are being filtered. Different kinds of filters are used, but cellulose acetate filters are probably most common. Most important is that the filters do not release any DOC during filtration. Samples from the mineral soil collected with tension lysimeters and samples from stream water may not always need filtering, but this needs to be evaluated for each site.

Numerous analyzers exist on the market, most of which are termed TOC analyzers. Measurement of DOC entails removing inorganic carbon with acid, sparging the resultant CO, and oxidizing the remaining C (presumably all OC) and measuring the CO, generated by the oxidation process. Oxidation of DOC can be accomplished by combustion, UV persulfate oxidation, ozone, or through UV fluorescence.

Calculation of Fluxes

DOC fluxes are simply calculated by multiplying DOC concentration by the water flux; however, at the plot scale, probably the most difficult measurement is the flux of water. Due to disturbed hydrology, it is usually not possible to use the amounts of water collected in the lysimeters to estimate the flux. Computer models are sometimes used to estimate water fluxes. The measurement of soil moisture and hydraulic conductivity is one method to estimate the flux of water through the rooting zone. Micrometeorological techniques can also be applied by measuring surface inputs of precipitation or throughfall, estimating surface outputs in the form of evapotranspiration through the energy balance and measuring changes in soil moisture. Transport through the rooting zone can be calculated by difference using the hydrological mass balance.

At the watershed scale, typically flow is either measured by a device such as a weir or flume that is at the outlet of the stream exiting the watershed. Where such devices are unavailable, stream gauging is commonly employed. Stream gauging entails measuring flow and relating flow to stream water height or stage height. Regression relationships are developed over a range of flows relating stage height to flow.

Measurement of water fluxes to lakes and wetlands typically entails the use of groundwater wells and piezometers that measure the head of water upslope of the water body which allows for the estimation of inputs in saturated soil zones if one knows soil hydraulic conductivity.

Comparison of DOC fluxes anlong ecosystems, treatments or over time can elucidate changes in ecosystem processes. The few studies assessing harvesting on stream DOC vary in the response. Harvesting generally increases soil temperatures but also reduces redox status. Studies that see increases relate DOC increases simply to flow increases. A number of studies have demonstrated that the amount of wetlands, especially peatlands, controls watershed level transport of DOC in streams. If there are wetlands present in the watershed, that factor appears to overwhelm any vegetation management factor controlling DOC transport. A number of watersheds have been experiencing increases in DOC transport as a result of increasing temperatures from climate change; however, other studies indicate that decreases in atmospheric deposition of sulfur may be the cause of the increases. As a result of changes in land use, management practices, climate and atmospheric inputs, DOC will continue to be an important response variable as we strive to understand carbon storage and fluxes.

Heavy Water

Water is a transparent and nearly colorless chemical substance that is the main constituent of Earth's ocean, streams, and lakes. We may have heard of heavy water and wondered how it was different from ordinary water. Heavy water is water that contains heavy hydrogen or deuterium. The different properties and uses of heavy water are discussed below.

Heavy water or deuterium oxide (D_2O) is a form of water that contains a large amount of the hydrogen isotope deuterium which is also known as heavy hydrogen. Deuterium differs from the hydrogen which is usually found in water. Heavy water may be deuterium protium oxide (DHO) or deuterium oxide (D_2O).

The increase in mass due to the presence of deuterium gives it a different chemical and physical property compared to normal water.

Some other Heavy Forms of water are:

- Tritiated water

- Heavy-oxygen water

- Semiheavy water

Properties of Heavy Water

Heavy water contains a huge amount of deuterium oxide. The properties of heavy water are mentioned in the table below:

Properties	H2O (Light water)	D2O (Heavy water)
Boiling point	100.0° C (212° F) (373.15 K)	101.4° C (214.5° F) (374.55 K)
Freezing point	0.0° C (32° F) (273.15 K)	3.82° C (38.88° F) (276.97 K)
Temp. of maximum density	3.98° C	11.6° C
Density at STP (g/mL)	0.9982	1.1056
Surface tension (at 25° C, N/m)	0.07198	0.07187
Dynamic viscosity (at 20° C, mPa·s)	1.0016	1.2467
Heat of vaporisation (kJ/mol)	40.657	41.521
Heat of fusion (kJ/mol)	6.00678	6.132
pK_b (at 25° C)	7.0	7.44 ("pKb D_2O")
pH (at 25° C)	7.0	7.44 ("pD")
Refractive index (at 20° C, 0.5893 μm)	1.33335	1.32844

Uses of Heavy Water

Heavy water is used in certain types of nuclear reactors, where it acts as a neutron moderator to slow down neutrons. The different applications and uses of heavy water are mentioned below:

- Nuclear magnetic resonance

- In Organic chemistry

- Fourier transform spectroscopy

- Neutron moderator

- Neutrino detector

- Metabolic rate testing in physiology and biology

- Tritium production

Light Water

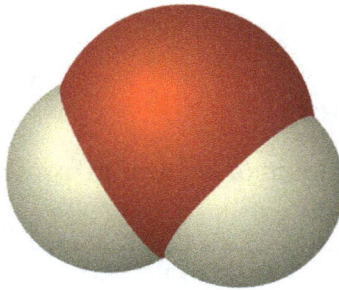

Figure: A regular water molecule.

Light water is simply ordinary water that does not contain large amounts of deuterium, making it distinct from heavy water. Although this water does contain small numbers of heavy water molecules, it isn't enough to make any significant changes in its properties. Light water plays an important role in the generation of electricity from nuclear energy, as it can serve both as a moderator and a coolant to carry away the energy generated by nuclear fission.

Use as a Moderator

In nuclear fission reactors, the neutrons must be slowed down to ensure an effective fission chain reaction occurs. This process of slowing neutrons down is known as moderation, and the material that slows down these neutrons is known as a neutron moderator. Light water can be used as a moderator in certain reactors, mainly pressurized water reactors and boiling water reactors. Light water can only work as a moderator in certain situations, as it absorbs too many neutrons to be used with uranium that is unenriched, so uranium enrichment is necessary to operate reactors that use light water as the moderator.This increases the overall cost of the operation, but makes light water reactors cheaper to build.

The simplest of these light water reactors is the boiling water reactor. More than 80% of the world's nuclear power plants use these light water reactors, with light water as their moderator. In light water reactors, there exists a thick-walled pressure vessel that contains the nuclear fuel and the moderator and coolant water circulates among the fuel rods to slow neutrons and carry away thermal energy.

There are several safety benefits that come from using light water as a moderator. First, the loss of any coolant deprives the reactor of its moderator, stopping the nuclear chain reaction. However, radioactive decay still continues to produce energy. As well, any increase in the temperature of the reactor causes the water to become less dense, reducing the level of moderation that the light water supplies and reducing the activity in the reactor. This means that if the reactivity increases too much, there will be less moderation to slow the nuclear reaction.

Doubly Labeled Water

Doubly labeled water is water in which both the hydrogen and the oxygen have been partly or completely replaced with an uncommon isotope of these elements for tracing purposes.

In practice, for both practical and safety reasons, almost all recent applications of the "doubly labeled water" method use water labeled with heavy but non-radioactive forms of each element. In theory, radioactive heavy isotopes of the elements could be used for such labeling; this was the case in many early applications of the method.

In particular, doubly labeled water can be used for a method to measure the average daily metabolic rate of an organism over a period of time. This is done by administering a dose of DLW, then measuring the elimination rates of deuterium and oxygen-18 in the subject over time (through regular sampling of heavy isotope concentrations in body water, by sampling saliva, urine, or blood). At least two samples are required: an initial sample (after the isotopes have reached equilibrium in the body), and a second sample some time later. The time between these samples depends on the size of the animal. In small animals, the period may be as short as 24 hours; in larger animals (such as adult humans), the period may be as long as 14 days.

The method was invented in the 1950s by Nathan Lifson and colleagues at the University of Minnesota. However, its use was restricted to small animals until the 1980s because of the high cost of the oxygen-18 isotope. Advances in mass spectrometry during the 1970s and early 1980s reduced the amount of isotope required, which made it feasible to apply the method to larger animals, including humans. The first application to humans was in 1982, by Dale Schoeller, over 25 years after the method was initially discovered. A complete summary of the technique is provided in a book by British biologist John Speakman.

Mechanism of the Test

The technique measures a subject's carbon dioxide production during the interval between first and last body water samples. The method depends on the details of carbon metabolism in our bodies. When cellular respiration breaks down carbon-containing molecules to release energy, carbon dioxide is released as a byproduct. Carbon dioxide contains two oxygen atoms and only one carbon atom, but food molecules such as carbohydrates do not contain enough oxygen to provide both oxygen atoms found in CO_2. It turns out, one of the two oxygen atoms in CO_2 is derived from body water. If the oxygen in water is labeled with ^{18}O, then CO_2 produced by respiration will contain labeled oxygen. In addition, as CO_2 travels from the site of respiration through the cytoplasm of a cell, through the interstitial fluids, into the bloodstream and then to the lungs some of it is reversibly converted to bicarbonate. So, after consuming water labeled with ^{18}O, the ^{18}O equilibrates with the body's bicarbonate and dissolved carbon dioxide pool. As carbon dioxide is exhaled, ^{18}O is lost from the body. This was discovered by Lifson in 1949. However, ^{18}O is also lost through body water loss. However, deuterium is lost *only* when body water is lost. Thus the loss of deuterium in body water over time can be used to mathematically compensate for the loss of ^{18}O by the water-loss route. This leaves only the remaining net loss of ^{18}O in carbon dioxide. This measurement of the amount of carbon dioxide lost is an excellent estimate for total carbon dioxide production. Once this is known, the total metabolic rate may be estimated from simplifying assumptions regarding the ratio of oxygen used in metabolism, to carbon dioxide eliminated. This quotient can be measured in other ways, and almost always has a value between 0.7 and 1.0, and for a mixed diet is usually about 0.8.

In lay terms:

- Metabolism can be calculated from oxygen-in/CO_2-out.

- DLW ('tagged') water is traceable hydrogen (deuterium), and traceable oxygen (^{18}O).

- The ^{18}O leaves the body in two ways: (i) exhaled CO_2, and (ii) water loss in (mostly) urine, sweat, & breath.

- BUT: the deuterium leaves *only* in the second way.

SO: from deuterium loss, we know how much of the tagged water left the body *as* water. And, since the concentration of ^{18}O in the body's water is measured after the labeling dose is given, we *also* know how much of the tagged oxygen left the body in the water. Measurement of ^{18}O dilution with time gives the total loss of this isotope by all routes (by water and respiration). Since the ratio of ^{18}O to total water oxygen in the body is measured, we can convert ^{18}O loss in respiration to total oxygen lost from the body's water pool via conversion to carbon dioxide. How much oxygen left the body *as* CO_2 is the same as the CO_2 produced by metabolism, since the body only produces CO_2 by this route. The CO_2 loss tells us the energy produced, if we know or can estimate the respiratory quotient.

Practical Isotope Administration

Doubly labeled water may be administered by injection, or orally. Since the isotopes will be diluted in body water, there is no need to administer them in a state of high isotopic purity, no need to employ water in which all or even most atoms are heavy atoms, or even to begin with water which is doubly labeled. Nor is it necessary to administer exactly one atom of ^{18}O for every two atoms of deuterium. This matter in practice is governed by the economics of buying ^{18}O enriched water, and the sensitivity of the mass-spectrographic equipment available.

In practice, doses of doubly labeled water for metabolic work are prepared by simply mixing a dose of deuterium oxide with a second dose of $H_2{}^{18}O$, which is water which has been separately enriched with ^{18}O, but otherwise contains normal hydrogen. The mixed water sample then contains both types of heavy atoms, in a far higher degree than normal water, and is now "doubly labeled." The free interchange of hydrogens between water molecules in liquid water ensures that the pools of oxygen and hydrogen in any sample of water will be separately equilibrated in a short time with any dose of added heavy isotope(s).

Applications

The doubly labeled water method is particularly useful for measuring average metabolic rate over relatively long periods of time, in subjects for which other types of direct or indirect calorimetric measurements of metabolic rate would be difficult or impossible. For example, the technique can measure the metabolism of animals in the wild state, with the technical problems being related mainly to how to administer the dose of isotope, and collect several samples of body water at later times to check for differential isotope elimination.

Most animal studies involve capturing the subject animals and injecting them, then holding them for a variable period before the first blood sample has been collected. This period depends on the size of the animal involved and varies between 30 minutes for very small animals to 6 hours for much larger animals. In both animals and humans, the test is made more accurate if a single determination of respiratory quotient has been made for the organism eating the standard diet at the time of measurement, since this value changes relatively little compared with the much larger metabolic rate changes related to thermoregulation and activity.

Because the heavy hydrogen and oxygen isotopes used in the standard doubly labeled water measurement are non-radioactive, and also non-toxic in the doses used, the doubly labeled water measurement of mean metabolic rate has been used extensively in human volunteers, and even in infants and pregnant women. The technique has been used on over 200 species of wild animals. Applications of the method to animals have been reviewed.

Hydronium

The hydrogen ion in aqueous solution is no more than a proton, a bare nucleus. Although it carries only a single unit of positive charge, this charge is concentrated into a volume of space that is only about a hundred-millionth as large as the volume occupied by the smallest atom. The resulting extraordinarily high charge density of the proton strongly attracts it to any part of a nearby atom or molecule in which there is an excess of negative charge. In the case of water, this will be the lone pair (unshared) electrons of the oxygen atom; the tiny proton will be buried within the lone pair and will form a shared-electron (coordinate) bond with it, creating a hydronium ion, H_3O^+. In a sense, H_2O is acting as a base here, and the product H_3O^+ is the conjugate acid of water:

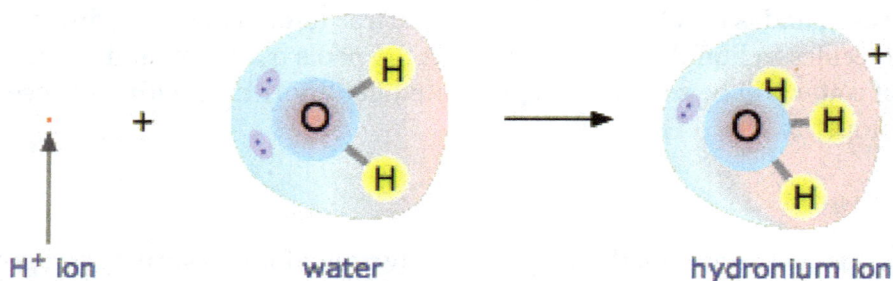

H⁺ ion   water   hydronium ion

Although other kinds of dissolved ions have water molecules bound to them more or less tightly, the interaction between H^+ and H_2O is so strong that writing "$H+_{(aq)}$" hardly does it justice, although it is formally correct. The formula H_3O^+ more adequately conveys the sense that it is both a molecule in its own right, and is also the conjugate acid of water.

The equation "$HA \rightarrow H^+ + A^-$" is so much easier to write that chemists still use it to represent acid-base reactions in contexts in which the proton donor-acceptor mechanism does not need to be emphasized. Thus, it is permissible to talk about "hydrogen ions" and use the formula H^+ in writing chemical equations as long as you remember that they are not to be taken literally in the context of aqueous solutions.

Interestingly, experiments indicate that the proton does not stick to a single H_2O molecule, but changes partners many times per second. This molecular promiscuity, a consequence of the uniquely small size and mass the proton, allows it to move through the solution by rapidly hopping from one H_2O molecule to the next, creating a new H_3O^+ ion as it goes. The overall effect is the same as if the H_3O^+ ion itself were moving. Similarly, a hydroxide ion, which can be considered to be a "proton hole" in the water, serves as a landing point for a proton from another H_2O molecule, so that the OH^- ion hops about in the same way.

Because hydronium and hydroxide ions can "move without actually moving" and thus without having to plow their way through the solution by shoving aside water molecules, as do other ions, solutions which are acidic or alkaline have extraordinarily high electrical conductivities.

Reaction

The hydronium ion is an important factor when dealing with chemical reactions that occur in aqueous solutions. Its concentration relative to hydroxide is a direct measure of the pH of a solution. It can be formed when an acid is present in water or simply in pure water. It's chemical formula is H_3O^+. It can also be formed by the combination of a H^+ ion with an H_2O molecule. The hydronium ion has a trigonal pyramidal geometry and is composed of three hydrogen atoms and one oxygen atom. There is a lone pair of electrons on the oxygen giving it this shape. The bond angle between the atoms is 113 degrees.

$$H_2O_{(l)} \rightleftharpoons OH^-_{(aq)} + H^+_{(aq)}$$

As H^+ ions are formed, they bond with H_2O molecules in the solution to form H_3O^+ (the hydronium ion). This is because hydrogen ions do not exist in aqueous solutions, but take the form of the hydronium ion, H_3O^+. A reversible reaction is one in which the reaction goes both ways. In other words, the water molecules dissociate while the OH$^-$ ions combine with the H^+ ions to form water. Water has the ability to attract H^+ ions because it is a polar molecule. This means that it has a partial charge, in this case the charge is negative. The partial charge is caused by the fact that oxygen is more electronegative than hydrogen. This means that in the bond between hydrogen and oxygen, oxygen "pulls" harder on the shared electrons thus causing a partial negative charge on the molecule and causing it to be attracted to the positive charge of H^+ to form hydronium. Another way to describe why the water molecule is considered polar is through the concept of dipole moment. The electron geometry of water is tetrahedral and the molecular geometry is bent. This bent geometry is asymmetrical, which causes the molecule to be polar and have a dipole moment, resulting in a partial charge.

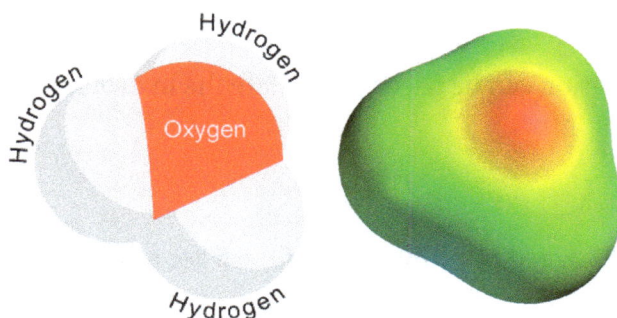

Figure: The picture above illustrates the electron density of hydronium.

The red area represents oxygen; this is the area where the electrostatic potential is the highest and the electrons are most dense.

An overall reaction for the dissociation of water to form hydronium can be seen here:

$$2H_2O_{(l)} \rightleftharpoons OH^-_{(aq)} + H_3O^+_{(aq)}$$

With Acids

Hydronium not only forms as a result of the dissociation of water, but also forms when water is in the presence of an acid. As the acid dissociates, the H^+ ions bond with water molecules to form hydronium, as seen here when hydrochloric acid is in the presence of water:

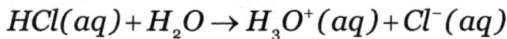

$$HCl(aq) + H_2O \rightarrow H_3O^+(aq) + Cl^-(aq)$$

pH

The pH of a solution depends on its hydronium concentration. In a sample of pure water, the hydronium concentration is 1×10^{-7} moles per liter (0.0000001 M) at room temperature. The equation to find the pH of a solution using its hydronium concentration is:

$$pH = -log(H_3O^+)$$

Using this equation, we find the pH of pure water to be 7. This is considered to be neutral on the pH scale. The pH can either go up or down depending on the change in hydronium concentration. If the hydronium concentration increases, the pH decreases, causing the solution to become more acidic. This happens when an acid is introduced. As H^+ ions dissociate from the acid and bond with water, they form hydronium ions, thus increasing the hydronium concentration of the solution. If the hydronium concentration decreases, the pH increases, resulting in a solution that is less acidic and more basic. This is caused by the OH^- ions that dissociate from bases. These ions bond with H^+ ions from the dissociation of water to form H_2O rather than hydronium ions.

A variation of the equation can be used to calculate the hydronium concentration when a pH is given to us:

$$H_3O^+ = 10^{-pH}$$

When the pH of 7 is plugged into this equation, we get a concentration of 0.0000001 M as we should.

Water Cluster

It is clear that life on Earth depends on the unusual structure and anomalous nature of liquid water. Organisms consist mostly of liquid water. This water performs many functions and it can never be considered simply as an inert diluent; it transports, lubricates, reacts, stabilizes, signals, structures, and partitions. The living world should be thought of as an equal partnership between the biological molecules and water. Both short-range (< 1 nm) and long range (> 100 nm) organization of the water molecules has been detected.

In spite of much work, many of the properties of water are puzzling. Enlightenment comes from an understanding that water molecules form an infinite dynamic hydrogen-bonded network with localized and structured clustering. The middling strength of the connecting hydrogen bonds seems ideally suited to life processes, being easily formed but not too difficult to break. An important concept, often overlooked, is that althoiugh liquid water seems homogeneous at macroscopic length scales and time scales, it is not homogeneous at the nanoscopic time or length scales.

Two water tetramer clusters forming an octamer cluster:

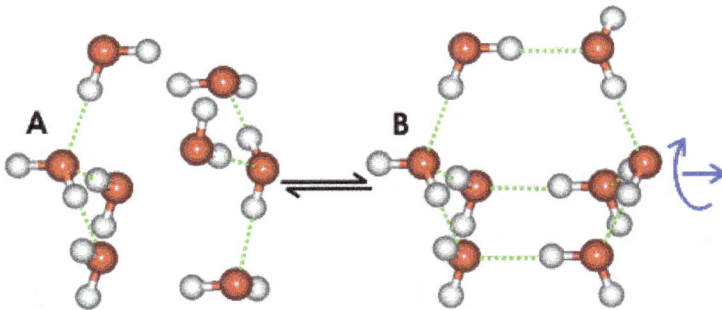

Small clusters of four water molecules may come together to form water bicyclo-octamers. The molecular arrangement (A) also occurs in high-density ice-seven whereas, with 60° relative twist, (B) is found in low-density hexagonal ice. Structures similar to A have greater numbers of 3-hydrogen-bonded and 5-coordinated water molecules as found at higher temperatures in liquid water, whereas structures similar to B have greater numbers of 4-hydrogen-bonded and 4-coordinated water molecules as found at lower temperatures in liquid water. Such equilibria are balanced due to the existence of two minima in the potential energy (U) - molecular separation (r) diagram below, which shows the approach of the water tetramers. It has been found using recent high energy x-ray diffraction experiments that the number of water nearest neighbors does not vary over the range 254.2 – 365.9 K with neighboring molecules switching between strong hydrogen-bonded and weak or non-hydrogen-bonded, in agreement with this model.

This competition between maximizing Vander Waals interactions (A, yielding higher orientation entropy, higher density and individually weaker but more numerous water-water binding energies) and maximizing hydrogen bonding (B, yielding more ordered structuring, lower density, and fewer but stronger water-water binding energies) is finely balanced, easily shifted with changed physical conditions, solutes, and surfaces. The potential energy barrier between these states ensures that water molecules prefer either structure A or B with little time spent on intermediate structures. An individual water molecule may be in state A with respect to some neighbors whilst being in state B with respect to others (for example, ice-seven).

Certainly, recent simulations using ab initio van der Waals interactions support this mechanism for the density fluctuations in liquid water.

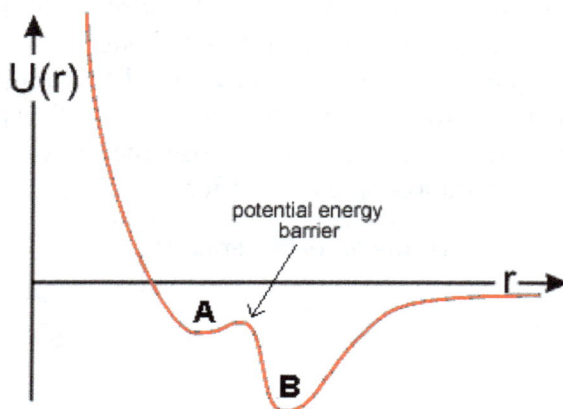

Figure: Potential energy diagram of the approach of water tetramers showing a shallow minimum inside a deeper minimum.

The shallow minimum (a), due to non-bonded interactions, lies up to 20% inside the deeper minimum (b) due to hydrogen bonding (even allowing for a 15% closer approach of individual hydrogen-bonded water molecules). In spatial terms, minimum (a) is far more extensive as the hydrogen-bonded minimum (b) is restricted in its geometry, being highly directional. At lower temperatures (particularly below the temperature of maximum density) and pressures, the less dense structure with more extensive hydrogen bonding at the lower minimum (b) will be preferred even though it involves a more ordered (lower entropy) structure. At higher temperatures, non-bonded interactions dominate causing breakdown of the clustering.

The hydrogen bonding, although cohesive in nature, is thus holding the water molecules apart. It is the conflict between these two effects, and how it varies with conditions, which endows water with many of its unusual properties. These bicyclo-octamers may cluster further, with only themselves, to form highly symmetric 280-molecule icosahedral water clusters that are able to interlink and tessellate throughout space. A mixture of water cyclic pentamers and tricyclo-decamers can bring about the same resultant clustering.

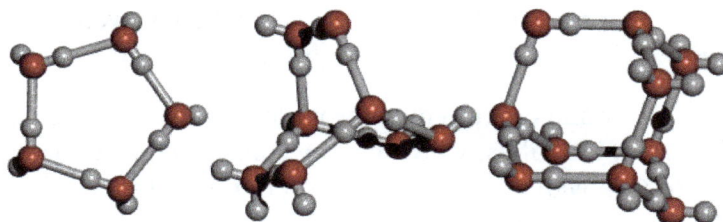

Cyclic pentamer   Bicyclo-octamer   Tricyclo-decamer

As all three of these small clusters are relatively stable quasi-polyhedra found, it is likely that their interaction will produce these larger icosahedral clusters (and the latter two are the lowest energy structures out of those found by simulating supercooled water). Such clusters can dynamically form a continuous network of both open, low-density, and condensed structures.

It is important to recognize that whenever there is a cluster of water molecules in liquid water, there will be a large number of 'decorating' water molecules on their periphery. Thus the linear H_2O chain, cyclic pentamer, bicyclo-octamer, tricyclo-decamer, $(H_2O)20$ dodecahedron, $(H_2O)100$, $(H_2O)280$ (ES) clusters form only $\approx 33\%$, $\approx 33\%$, $\approx 36\%$, $\approx 38\%$, $\approx 50\%$, $\approx 63\%$, $\approx 70\%$ respectively of their cluster+'decorating' H_2O assemblage. These 'decorating' H_2O molecules necessarily have different properties than the internal clusters and naturally form a second 'state' of water molecule present.

Independence of Cluster Lifetimes and Hydrogen-bond Lifetimes

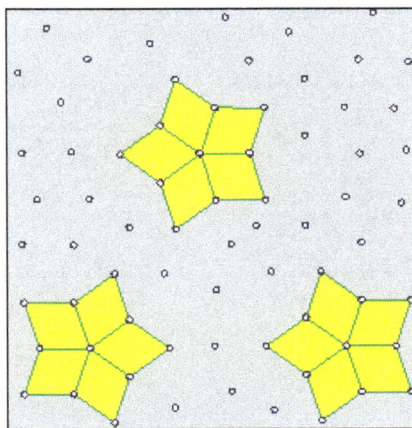

The figure gives the understanding of how the lifetimes of clusters are independent of the lifetime of individual linkages. The actual clusters of water molecules are not represented. It is supposed that the star clusters may reform around key structures. For each shifting cluster, a few units move to break up the existing cluster and help create a new cluster. The new clusters are identical to the old ones but only contain a proportion of the units. Thus dynamic clusters may reform around any of the star arms. One mechanism for such cluster reorientation involves a proton cascade leading to cluster reorientation. Movement of such clusters may involve them traveling as a wave through the medium as well as flickering on and off unpredictably.

Although the hydrogen bonds between water molecules in ice have equally short life-times to those in water, ice cubes, which can be considered as enormous ice clusters, can last forever at 0° C in water. As there is a shift of the water O-H band center of Raman scattering ($\approx$ 3300 cm-1) to higher frequencies (shorter wavenumbers, $\approx$ 40 cm-1) with decreasing the probe pulse duration from 20 ns to ~50 ps, it is indicated that there is an increased time required for the formation of large clusters in water via hydrogen bonding between H_2O molecules.

Similarly but on a macro-scale on the addition of solutes, water retains its integrity as liquid water with the hydrogen bond network connected throughout the entire bulk and yet the local hydrogen bonds are known to be fleetingly breaking and forming. Cluster formation and lifetime are both increased by the highly cooperative nature of hydrogen bonding. Whenever low-density clusters are formed they are surrounded by about the same number of 'decorating' water molecules with intermediate density.

There is a similarity, in principle, with John Conway's game "life" in the persistence of some of its structures.

Icosahedral Water Clusters

Cluster equilibrium, showing how the expanded low-density icosahedral cluster $(H_2O)280$ undergoes a partial collapse to give a condensed structure followed by a further fast collapse

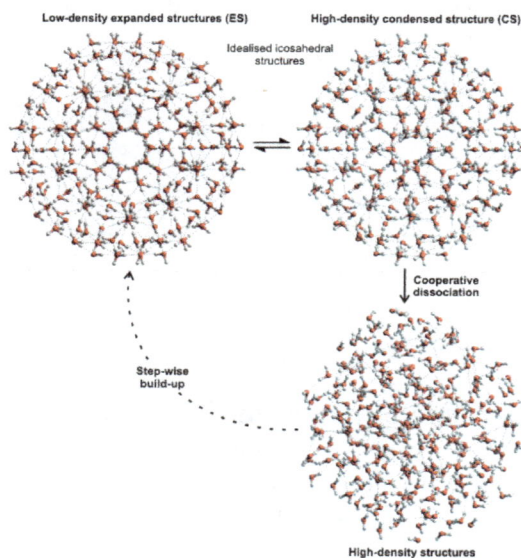

Such a dynamic fluctuating self-replicating network of water molecules, with localized and overlapping icosahedral symmetry, was first proposed to exist in liquid water in

1998 and the structure subsequently independently found, by X-ray diffraction, in water nanodrops in 2001. The clusters formed can interconvert between lower and higher density forms by bending, but not breaking, some of the hydrogen bonds. Structuring may also flicker between statistically and topographically equivalent clusters but involving different molecules by shifting their cluster centers. These polyhedral structures are idealized in the diagrams but are considerably distorted and fragmented by thermal effects in reality. The existence of long-lived ring fragments is nevertheless considered to be well-founded. The cluster size required for ice formation has been estimated at about 400 molecules (one further layer of water around the icosahedral water core), although the structure of this core structure was indeterminable. As the temperature increases the average cluster size, the cluster integrity and the proportion in the low-density form all decrease. This structuring accommodates explanation of many of the anomalous properties of water including its temperature-density and pressure-viscosity behavior, the radial distribution pattern, the presence of both cyclic pentamers and hexamers, the change in properties on supercooling a and the solvation and hydration properties of ions, hydrophobic molecules, carbohydrates, and macromolecules. The model described here offers a "two-state" structural model on to which large molecules can be mapped in order to offer insights into their interactions.

Lifetime of the Clusters

The open low-density cluster ES is seen as the the extreme form of water cluster with its subclusters building up as it is mostly formed in deeply supercooled water or by incorporating stabilising solutes. It is a relatively long-lived cluster both as a whole and in its parts, particularly the $(H_2O)20$ and $(H_2O)100$ subclusters. Its collapse into the condensed structure (CS) is espected to be a major route for its degradation (rather than slow loss from its exterior) that may be caused by increased pressure, temperature or destabilising solutes. This sudden extensive and comprehensive collapse is a key feature of the two-state structuring of water. The condensed structure is (CS) that is far less stable than the ES structure or its substructures. It can further change either by reforming ES or by a further abrupt and comprehensive collpase to form high density stranded and small ring structures.

Water Dimer

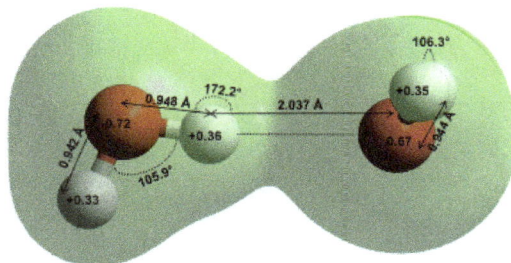

Figure: Water Dimer

Much effort has been expended on the structure of small isolated water clusters. The smallest cluster is the dimer. Typically, in the ambient atmosphere, there is over one water dimer for every thousand free water molecules rising to about one in twenty in steam. It has one mirror plane of symmetry (CS) and dimensions from ab initio 6-31G** calculation as shown right. The charge transfer for this calculation is only 0.03 e⁻ from δ+ve acceptor molecule to δ-ve donor molecule. Its molecular orbitals are shown on another page.

The Variation of Water's Fugacity Coefficient with Pressure

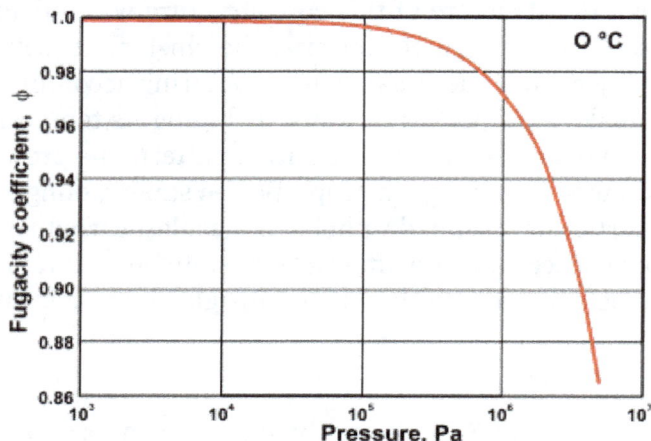

The equilibrium constant for dimer formation is 0.0501 bar^{-1} at 298.15 K, with dissociation energy of 13.2 kJ × mol^{-1}, $(D_2O)_2$ 14.9 kJ × mol^{-1}. The thermochemical properties of the dimer have been determined and experimental dimer studies have been reviewed.

$$2H_2O(g) \rightarrow (H_2O)^2(g)$$

The dimer formation causes deviations from ideal behavior in gaseous water.

The effect of the hydrogen bond on the stretch vibrations is seen below. The activation energy required to switch from one acceptor position to the other on the acceptor molecule (1.88 kJ × mol^{-1}) is only 40% of that required to switch from one donor hydrogen to the other (4.71 kJ × mol^{-1}) on the donor molecule via the bifurcated state, both of course far lower than the dissociation energy.

Vibrational bands	symmetric stretch $(v_1,\ cm^{-1})$	bend $(v_2,\ cm^{-1})$	asymmetric stretch $(v_3,\ cm^{-1})$
H_2O monomer	3657	1595	3756
HO-H···OH$_2$ dimer donor	3545	1669	3715
HO-H···OH$_2$ dimer acceptor	3600	1653	3730

The high-resolution spectra for the out-of-plane librational vibrations, at around 500 cm^{-1}, of the water dimer has been published as has the rotational spectra around 4 cm^{-1}.

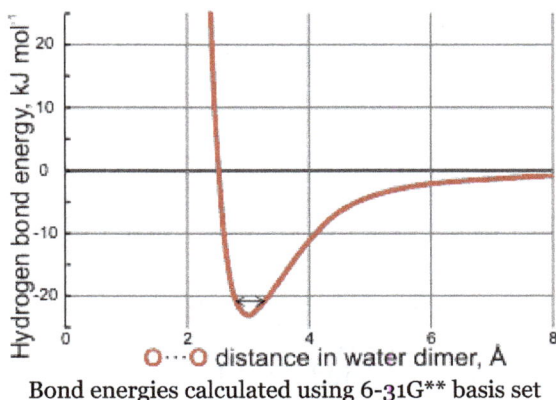

Bond energies calculated using 6-31G** basis set

Calculated bond energies for the water dimer are given right using the 6-31G** basis set. Particularly noteworthy is the steepness of the 'wall' inside the optimum position and the more relaxed structures allowed outside the optimum position that still have significant bond energy. The position of the hydrogen bond stretch vibration energy ($\approx$ 200 cm^{-1}) is shown with the vibrational range of about -0.15 Å +0.3 Å. Although stated in the early literature that the behavior is purely electrostatic at larger distances, this is not true. The most energetically favorable water dimer is shown above right using ab initio calculations with the 6-31G** basis set. It is also shown below with a section through the electron density distribution (high densities around the oxygen atoms have been omitted for clarity). This shows the tetrahedrality of the bonding in spite of the lack of clearly seen lone pair electrons; although a small amount of distortion along the hydrogen bond can be seen. This tetrahedrality is primarily caused by electrostatic effects (that is, repulsion between the positively charged non-bonded hydrogen atoms) rather than the presence of tetrahedrally placed lone pair electrons. The hydrogen-bonded proton has reduced electron density relative to the other protons. Note that, even at temperatures as low as a few kelvin, there are considerable oscillations in the hydrogen bond length ($\approx$ ±0.2 Å at 50 K with timescales $\approx$ 0.2 ps) and angles ($\approx$ 9 ° at 50 K with timescales $\approx$ 0.2 ps). The potential energy surface and wagging vibration of the water dimer have been described.

Water dimer showing the electron density perturbation along the hydrogen bond

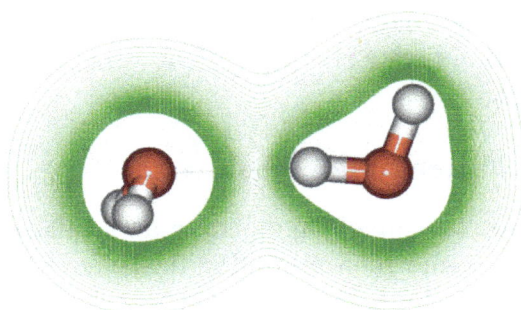

Water Dimer Dimensions

$R = 2.976$ (+0.000, -0.030) Å, $\alpha = 6 \pm 20°$, $\beta = 57 \pm 10°$, α is the donor angle and β is the acceptor angle. The dimer (with slightly different geometry) dipole moment is 2.6 D. Although β is close to as expected if the lone pair electrons were tetrahedrally placed (= 109.47°/2), the energy minimum ($\approx$ 21 kJ mol^{-1}) is broad and extends towards $\beta = 0°$.

It has been noted that dimers of more distant water molecules ($\approx$ 1 nm) show synchronous behavior due to their interacting electric fields.

Bond Energies of Dimers with Temperature.

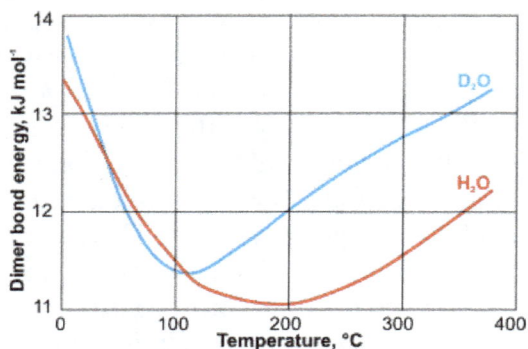

At lower temperatures, the bond energy for $(H_2O)_2$ and $(D_2O)_2$ dimers reduce with temperature, explained as due to the vibration amplitude growth. However at higher temperatures, the bond energies of both increase with temperature. This phenomenon also occurs in methanol but is difficult to explain.

There are two HOD dimer isomers, the D-bonded isomer and the H-bonded isomer. The H-bonded isomer ground state energy is 686 J × mol^{-1} higher than the D-bonded isomer ground state energy.

Water Dimer and Trimer Inside Fullerenes

Using clever synthetic methods involving opening up a fullerene, inserting the water molecule(s) under pressure and then re-sealing the fullerene, one molecule of water has been placed in C_{60} and C_{70} fullerenes and two water molecules placed inside the C_{70} fullerene.

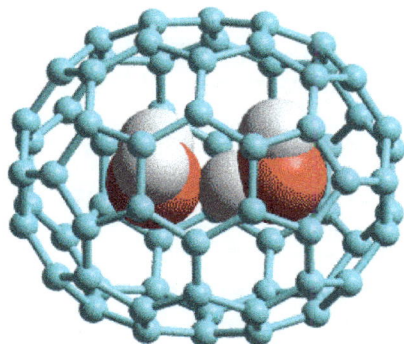

C$_{70}$ fullerene containing a water dimer

The water dimer inside the C$_{70}$ fullerene is free from any further hydrogen bonding to water molecules and is prevented from dissociating due to the confinement. It has a cis-linear conformation resulting from confinement (3.7 Å x 4.6 Å prolate ellipsoid cavity) and high effective pressure effects inside C$_{70}$. This has a structure similar to the most energetically favorable water trans dimer with the important change in that the water molecule on the left is flipped up rather than down.

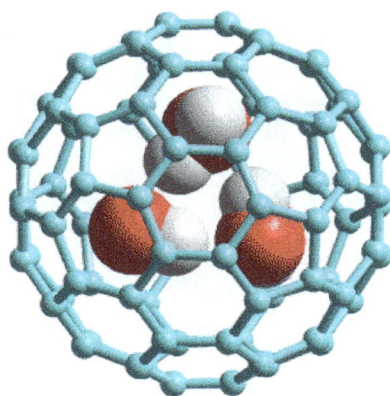

C$_{84}$ fullerene containing a water trimer

The cyclic water-trimer is predicted to be able to be placed into the D$_2$(22)-C84 fullerene (one of 24 isomers) with a potential-energy gain of 43.5 kJ × mol^{-1}

Two different conformational forms are predicted to exist with the arrangement containing the three non-hydrogen bonded H atoms on the same side of the O-O-O plane (cis-organization) slightly more stable (0.3 kJ × mol^{-1} C$_{84}$) than the one similar (but not identical) to the gas-phase form (trans-organization), as shown left (two down one up). The ratio of these two forms is about 57%:43%. The hydrogen bonds are calculated to be slightly shorter than that in the water dimer inside the C$_{70}$ fullerene shown above right.

Water confined in larger fullerenes are calculated to exist in single to multiple concentric, spherical shells as the size of the fullerene increases with these water-cluster shells exhibiting solid-like behavior at temperatures as high as 500 K.

The spherical C_{180} fullerene can enclose the $(H_2O)_{20}$ dodecahedral cage, and the spherical C_{500} fullerene can enclose the $(H_2O)_{100}$ inner ES core but their structures are more determined by van der Waals interactions with the carbon cages than by hydrogen bonds.

The Water Trimer, Tetramer and Pentamer

The water trimer ring system has been reviewed up to 2003 and reanalyzed in 2016. It appears that quantum delocalization of hydrogen-bonded protons between oxygen neighbors occurs in the trimer and pentamer, but not the tetramer and hexamer. Such aromatic-like delocalization within pentamers and dodecahedra present in supercooled water stabilize their structures and contribute to the stability of ES-clustering, where half the water molecules are within pentamers.

The average dimensions for the trimer, tetramer, and pentamer from ab initio 6-31G** calculation are shown below. The charge on the donor hydrogen atoms increases the hydrogen bond lengths contract and the electron density width within the hydrogen bonds increase as the structure goes from dimer to trimer to tetramer to pentamer.

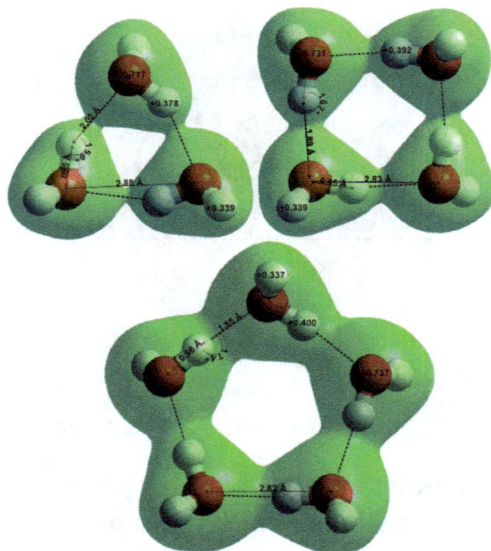

Structures of water trimer, tetramer, and pentamer, from ab initio 6-31G** calculation

Small Clusters

Many theoretical studies in vacuo and some experimental work in the gas phase have been carried out on small water clusters. Often the underlying theme is that the structures will be of use in understanding the structure of liquid water. This, however, is misleading as the lowest energy clusters, with greater than five water molecules, do not occur in liquid water. The reason for this is that the small water clusters described balance the strength of their hydrogen bonding against the number of hydrogen bonds that can be formed, often maximizing the number of hydrogen bonds with each hydrogen bond being of sub-maximal strength (17-19 kJ × mol⁻¹) and with somewhat strained angles (poor directionality). If such clusters were to be immersed in liquid water, many new hydrogen bonds would be established around their periphery with the surrounding water molecules in the liquid; with the result that the original cluster would immediately 'dissolve' and change its structure.

The structures of these small water clusters are interesting, however. The most stable structures with up to five water molecules are given as the above planar structures.

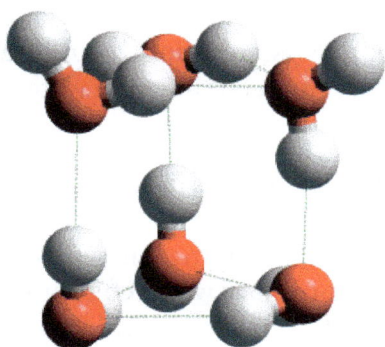

$(H_2O)_6$ prism with 9 hydrogen bonds $(H_2O)_8$ cube with 12 hydrogen bonds

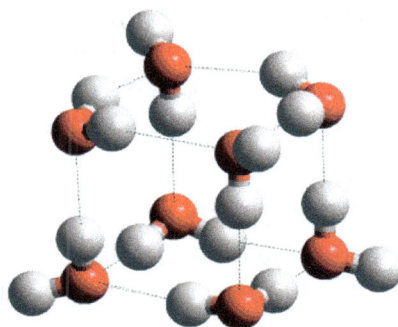

In liquid water, the preferred $(H_2O)_6$ conformation would be the chair hexamer with just 6 internal hydrogen bonds but with an additional 12 hydrogen bonds to other water molecules in the liquid (about twice as many and stronger hydrogen bonds than in the gas phase). Also in liquid water the preferred $(H_2O)_8$ conformation would be the bicyclo- octamer with just 9 internal hydrogen bonds but with an additional 14 hydrogen bonds to other water molecules in the liquid (also with about twice as many and stronger hydrogen bonds than in the gas phase). Larger structures have been found by ab initio calculation. As an example, $(H_2O)_{16}$ was shown to be optimal in a tri-stacked cube conformation with 28 hydrogen bonds; a structure that is highly unlikely to survive in a liquid water milieu.

References

- Speakman, J.R., Doubly Labelled Water: Theory and Practice. Springer Scientific publishers. ISBN 0-412-63780-4 ISBN 978-0412637803

- Water-molecule-structure, chemical-engineering: idconline.com, Retrieved 14 July 2018

- Jones PJ, Winthrop AL, Schoeller DA, et al. (March 1987). "Validation of doubly labeled water for assessing energy expenditure in infants". Pediatr. Res. 21 (3): 242–6. doi:10.1203/00006450-198703000-00007. PMID 3104873

- Vienna-Standard-Mean-Ocean-Water: wikidoc.org, Retrieved 27 June 2018

- Nagy, KA (2005) Field metabolic rates and body size. Journal of Experimental Biology 208, 1621–1625

- Heavy-water, chemistry: byjus.com, Retrieved 29 April 2018

4

Fundamental Concepts of Water Chemistry

The properties and chemical components of water are studied under water chemistry. An extensive study of the fundamental concepts and principles of water chemistry, such as metal aquo complexes, hydrolysis, water of crystallization, hydration reaction, electrolysis, dehydration reaction, etc. is vital to the understanding of water chemistry which have been covered in elaborate detail in this chapter.

Water is an unusual compound with unique physical properties. As a result, its the compound of life. Yet, its the most abundant compound in the biosphere of Earth. These properties are related to its electronic structure, bonding, and chemistry. However, due to its affinity for a variety of substances, ordinary water contains other substances. Few of us has used, seen or tested pure water, based on which we discuss its chemistry.

The chemistry of water deals with the fundamental chemical property and information about water. Water chemistry is discussed in the following subtitles:

- Composition of water
- Structure and bonding of water
- Molecular Vibration of water
- Symmetry of water molecules
- Formation of hydrogen bonding in water
- Structure of ice

- Autoionization

- Leveling effect of water and acid-base characters

- Amphiprotic nature

- Reactivity of water towards alkali metals; alkaline earth metals; halogens; hydrides; methane; oxides; and oxygen ions.

- Electrolysis of water

Composition of Water

Water consists of only hydrogen and oxygen. Both elements have natural stable and radioactive isotopes. Due to these isotopes, water molecules of masses roughly 18 ($H_2{}^{16}O$) to 22 ($D_2{}^{18}O$) are expected to form. Isotopes and their abundances of H and O are given below. From these data, we can estimate the relative abundances of all isotopic water molecules.

Abundances (% or halflife) of hydrogen and oxygen isotopes					
H	^2D	^3T			
99.985%	0.015%	12.33 y			
^{14}O	^{15}O	^{16}O	^{17}O	^{18}O	
70.6 s	122 s	99.762%	0.038%	0.200%	
Relative abundance of isotopic water					
$H_2{}^{16}O$	$H_2{}^{18}O$	$H_2{}^{17}O$	$HD^{16}O$	$D_2{}^{16}O$	$HT^{16}O$
99.78%	0.20%	0.03%	0.0149%	0.022 ppm	trace
18	20	19	19	20	20 amu

The predominant water molecules $H_2{}^{16}O$ have a mass of 18 amu, but molecules with mass 19 and 20 occur significantly. Because the isotopic abundances are not always the same due to their astronomical origin, The isotopic distribution of water molecules depends on its source and age. Its study is linked to other sciences.

In particular, $D_2{}^{16}O$ is called heavy water, and it is produced by enrichment from natural water. Properties of heavy water are particularly interesting due to its application in nuclear technology.

Structure and Bonding of the Water Molecule

Pure water, H_2O, has a unique molecular structure. The O-H bondlengths are 0.096 nm and the H-O-H angle = 104.5°. This strange geometry can be explained by various methods.

Lewis Dot Structures

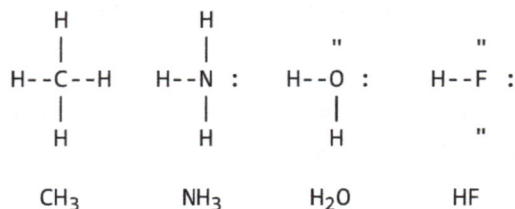

```
    H                H
    |                |              "              "
H--C--H          H--N :        H--O :         H--F :
    |                |              |
    H                H              H              "

   CH₃              NH₃           H₂O            HF
```

Bondlength/pm

C-H	N-H	O-H	H-F
109	101	96	92

From carbon to neon, the numbers of valence electrons increase from 4 to 8. These elements require 4, 3, 2, 1, and OH atoms to share electrons in order to complete the octet requirement. Their Lewis dot structures are shown on the right, and note the trend in bondlengths.

There are six valance electrons on the oxygen, and one each from the hydrogen atom in the water molecule. The eight electrons form two H-O bonds, and left two lone pairs. The long pairs and bonds stay away from each other and they extend towards the corners of a tetrahedron. Such an ideal structure should give H-O-H bond angle of 109.5°, but the lone pairs repel each other more than they repel the O-H bonds. Thus, the O-H bonds are pushed closer, making the H-O-H angle less than 109°.

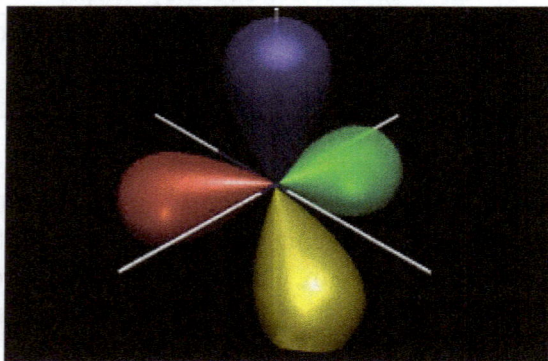

After the introduction of quantum mechanics, the electronic configuration for the valence electron of oxygen are $2s^2 2p^4$. Since the energy levels of $2s$ and $2p$ are close, valence electrons have characters of both s and p. The mixture is called sp^3 hybridization. These hybridized orbitals are shown on the right. The structures of CH_4, NH_3, and H_2O can all explained by these hybrid orbitals of the central atoms. The above approach is the valence bond theory, and both the C-H bonds and lone electron pairs are counted as VSPER pairs in the Valence-shell Electron-Pair Repulsion (VSEPR) model, according to which, the four groups point to the corners of a tetrahedron.

For triatomic molecules such as water, molecular orbital (MO) approach can also be applied to discuss the bonding. The result however is similar to the valence bond approach, but the MO theory gives the energy levels of the electron for further exploration.

Molecular Vibration of Water

Atoms in a molecule are never at rest, and for each type of molecule, there are some normal vibration modes. For the water molecule, the three normal modes of vibrations are symmetric stretching, bending and assymmetric stretching.

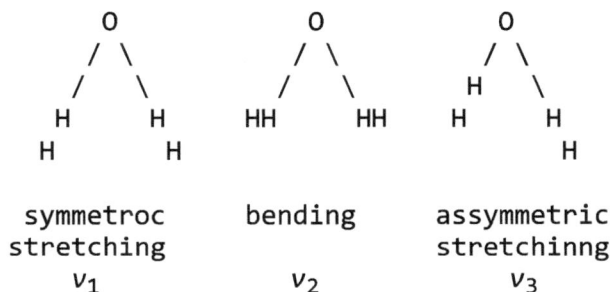

```
      O              O              O
     / \            / \            / \
    /   \          /   \         H    \
   H     H        HH     HH      H      H
   H     H                              H

 symmetroc         bending       assymmetric
 stretching                      stretchinng
    V1               V2              V3
```

Basic modes of vibration for H_2O

The vibrations are quantized, as do any microscopic system, and their quantum numbers are designated as v_1, v_2 and v_3. The observed transition bands of D_2O, H_2O, and HDO are given in the table on the right.

Transition bands of D_2O, H_2O, and HDO					
Quantum numbers of upper state			Absorption wavenumbers of bands /cm^{-1}		
v_1	v_2	v_3	D_2O	H_2O	HDO
0	1	0	1178	1594	1402
1	0	0	2671	3656	2726
0	0	1	2788	3756	3703
0	1	1	3956	5332	5089

The ideal transition bands are centered in the given wavenumbers. However, these wavenumbers are calculated based on isolated molecules with no interaction with any neighbour. When molecules interact with each other, the energy levels are modified, and the bands shift.

Many more less intense absorption bands extend into the green part of the visible spectrum. The absorption spectrum of water may contribute to the blue color for lake, river and ocean waters.

Symmetry of Water Molecules

The water molecules are rather symmetric in that there are two mirror planes of symmetry, one containing all three atoms and one perpendicular to the plane passing through the bisector of the H-O-H angle. Furthermore, if the molecules are rotated 180° (360°/2) the shape of the molecule is unperturbed. This indicates that the molecules have a 2-fold rotation axis. The three symmetry elements are 2-fold rotation, and two mirror planes. Both mirror planes contain the rotation axis, and this type of symmetry belongs to the point group C_{2v}.

A point group has a definite number of symmetry elements arranged in certain fashion. Molecules can be classified according to their point groups. Molecules of the same point group have similar spectroscopic characters. Other molecules of C_{2v} point group are $CH_2=O$, CH_2Cl_2, the bent O_3 etc.

```
        O
       / \
      H   H
```

Formation of Hydrogen Bonding

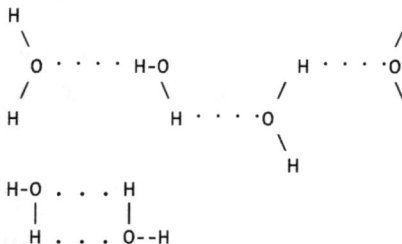

```
    H                                              /
     \                                            /
      O · · · · H-O            H · · · ·O
     /               \        /          \
    H                 H · · · ·O           H
                            \
                             H
    H-O . . . H
     |         |
     H . . . O--H
```

Dimer

Hydrogen bonds among water molecules.

Under certain conditions, an atom of hydrogen is attracted by rather strong forces to two atoms instead of only one, so that it may be considered to be acting as a bond between them. This is called hydrogen bond. This statement is from Linus Pauling (1939) in his book *The Nature of the Chemical Bond*. He gave the ion [F:H:F]⁻ as an example. At that time, the hydrogen bond was recognized as mainly ionic in nature. The energy associated with hydrogen bond is 8 to 40 kJ/mol.

Normally, the melting point and boiling point of a substance increase with molecular mass. For example the melting points of inert gases are 0.95, 24.48, 83.8, and 116.6 K respectively for He, Ne, Ar, and Kr.

In this table, the melting and boiling points for water are particular high for its small molecular mass. This is usually attributed to the formation of hydrogen bonds. The small electronegative atoms F, O and N are somewhat negatively charged when they are bonded to hydrogen atoms. The negative charges on F, O and N attract the slightly positive hydrogen atoms, forming a strong interaction called hydrogen bond.

Comparison of melting and boiling points for a few substances			
Molecule	Molar mass	m.p.	b.p. /° C
NH_3	17	-77.8	-33.5
H_2O	18	-0	100
H_2S	34	-85.6	-60
H_2Se	81	-60.4	-41.5
H_2Te	128.6	-51	-1.8
CH_3OH	32	?	65
C_2H_5OH	46	?	78
$C_2H_5OC_2H_5$	74	?	34

A graph showing the melting points and boiling points of group 16 provided by illustrates the same point.

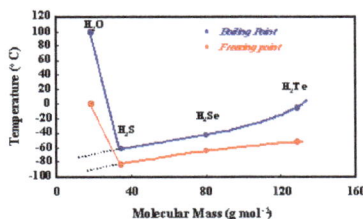

**Boiling and Freezing Points
of Group 16 Hydrides**

Based on the observed absorption at 3546 and 3691 cm^{-1}, Van Thiel, Becker, and Pinmentel suggested the formation of water dimer when trapped in a matrix of nitrogen.

Due to hydrogen bonding, water molecules form dimers, trimers, polymers, and clusters. The hydrogen bonds are not necessarily liner.

Structure of Ice

Ice occurs in many places, including the Antarctic. If all the ice melted, the water level of the oceans will rise about 70 m.

The density of ice is dramatically smaller than that of water, due to the regular arrangement of water molecule via hydrogen bonds. In an idealized structure of ice, every hydrogen atom is involved in hydrogen bond. Every oxygen atom is surrounded by four hydrogen bonds.

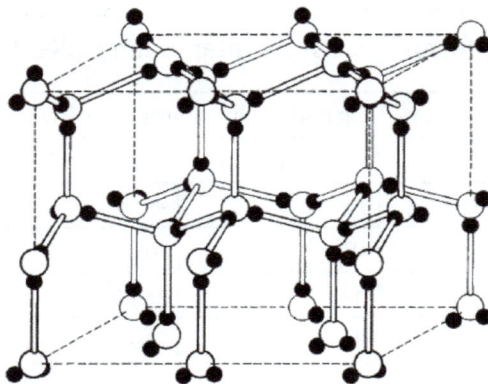

This diagram, shows the structure of hexagonal ice. Since the hydrogen bonds are not linear, the real structure is a little more complicated.

The tetrahedral coordination opens up the space between molecules. On each hydrogen bond, shown by a rod joining the oxygen atoms, lies one proton in an asymmetric position. Bond lengths, 275 pm, are indicated. Ordinary ice is hexagonal. and the hexagonal c axis is labelled 732 pm, and one of the hexagonal a axes is labelled 450 pm. If water vapor condenses on very cold substrate at 143-193 K (-130 to -80° C) a cubic phase is formed.

The rods joining the atoms represent C-C bonds. Each C-C bondlength is 154 pm. Silicon and germanium crystals have the same structure, but their bondlengths are longer. Hexagonal diamonds have been observed in meteorites.

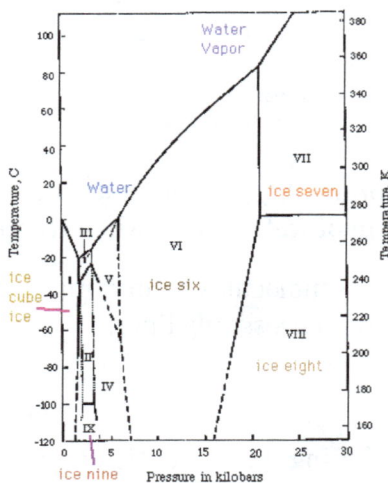

The four hydrogen bonds around an oxygen atom form a tetrahedron in a fashion found in the two types of diamonds. Thus, ice, diamond, and close packing of spheres are somewhat topologically related.

A phase diagram of water shows 9 different solid phases (ices). Ice Ih is the ordinary ice. In addition to ice Ic from vapor deposition, conditions for nine phases are shown. Aside from ice I, other phases are formed and observed under high pressure generated by machines built by scientists. So far, ten different forms of ice have been observed, and some ice forms exist at very high pressure. The pressure deep under the polar (Antarctic) ice cap is very high, but we are not able to make any direct observation or study.

There is a report of the 11th ice, and the ice phase diagram and drawings of ice structures given here is extremely interesting.

The Autoionization of Water

The Autoionization of Water in the formation of ions according to

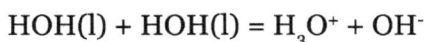

$$HOH(l) + HOH(l) = H_3O^+ + OH^-$$

This is an equilibrium process and is characterised by an equilibrium constant, K'_w:

$$K'_w = \frac{[H_3O^+]\,[OH^-]}{[H_2O]}$$

t° C	K_w
20	1.14e-15
25	1.00e-14
35	2.09e-14
40	2.92e-14
50	5.47e-14

Since $[H_2O] = 1000/18 = 55.56$ M, and remains rather constant under any circumstance, we usually write

$$K_w = [H_3O^+]\,[OH^-]$$
$$= 10^{-14} \text{ (or 1e-14)}$$
$$pK_w = -\log K_w \text{ (defined)}$$
$$= 14 \text{ (at 298 K)}$$

For neutral water, $[H_3O^+] = [OH^-] = $ 1e-7 at this temperature. Furthermore, we define

$$pH = -\log[H_3O^+]$$
$$pOH = -\log[OH^-]$$
$$pH = pOH = 7 \text{ at 298 K; (in neutral solutions)}$$

It is important to realize that K_w depends on temperature as shown in the Table here.

Leveling Effect of Water and Acid-base Characters

The strength of strong acids and bases is dominated by the autoionization of water. In aqueous solutions, the strongest acid and base are the hydronium ion, H_3O^+, and the hydroxide ion OH⁻ respectively. Acids HCl, HBr, HI, HNO_3, $HClO_3$, $HClO_4$, and H_2SO_4 completely ionize in water, making them as strong as H_3O^+ due to the leveling effect of water. Furthermore, strong acids, strong bases, and salts completely ionize in their aqueous solutions.

For example, HCl is a stronger acid than H_2O, and the reaction takes place as HCl dissolves in water.

$$HCl + H_2O = Cl^- + H_3O^+$$

A similar equation can be written for another strong acid.

On the other hand, a stong base also react with water to give the stong base species, OH⁻.

$$H_2O + B^- = OH^- + HB$$

For example, O^{2-}, CH_3O^-, and NH_3 are strong bases. The leveling effect also apply to bases.

Amphiprotic Species

Equilibria of acids and bases, are interesting chemistry. When an acid and a base differ by a proton, they are called a conjugate acid-base pair. A water molecule is a weak acid and base, due to its ability to accept or donate a proton. Such properties make water an amphiprotic species. In fact, H_3O^+, H_2O and OH⁻ are amphiprotic, as are some other conjugate acid-base pairs of weak acids and bases.

If several acids and bases are dissolved in water, all equilibria must be considered. To estimate the pH of these solutions requires the exact treatment of several equilibrium constants. For example, many species dissolve in rain water, and many equilibria must be considered. Detail consideration and examples are given in Acid-Base Reactions.

Carbon dioxide in the air dissolve in rain water, lakes and rivers. A solution of CO_2 involves the following reaction:

Reaction	K formula	K value
$H_2O(l) + CO_2(g) = H_2CO_3(l)$	$1/P_{CO2}$	?
$H_2CO_3 = HCO_3^- + H^+$	$[HCO_3^-][H^+]/[H_2CO_3]$	5e-7
$HCO_3^- = CO_3^{-2} + H^+$	$[CO_3^{-2}][H^+]/[HCO_3^-]$	5e-11
$HOH(l) + HOH(l) = H_3O^+ + OH^-$	$[H_3O^+][OH^-]$	1e-14

These complicated equilibria make natural water a buffer.

Example 1

Assume that the partial pressure of carbon dioxide causes a total concentration of carbonic species to be 8e-4 M. Estimate the pH of this solution.

Solution

From the given data, we have the following five equations and five unknowns:

Equilibrium   Equations   No.

$$H_2CO_3 \leftrightarrow HCO_3^- + H^+ \quad \frac{[HCO_3^-][H^+]}{[H_2CO_3]} = 5e\text{-}7 \quad (1)$$

$$HCO_3^- \leftrightarrow CO_3^{2-} + H^+ \quad \frac{[CO_3^{2-}][H^+]}{HCO_3^-} = 5e\text{-}11 \quad (2)$$

$$2\,H_2O \leftrightarrow H_3O^+ + OH^- \quad [H_3O^+][OH^-] = 1e\text{-}14 \quad (3)$$

Charge balance
$$[H^+] = [HCO_3^-] + [OH^-] + 2\,[CO_3^{2-}] \quad (4)$$

All species containing C $\dfrac{[H_2CO_3] + [HCO_3^-] + [CO_3^{2-}]}{}$ (5)
$$= 8.0e\text{-}4\ M$$

Unknown

$$[H^+], [OH^-], [H_2CO_3], [HCO_3^-], [CO_3^{2-}]$$

Solving these equations for the 5 unknowns can be done using Maple, Mathcad, spread sheet, or approximation. In any case, we are interested in the pH, and we can make the following approximations or assumptions

Assume H^+ mostly comes from (1) $[H^+] = [HCO_3^-]$

H_2CO_3 is a weak acid most unionize $[H_2CO_3] = 8.0e\text{-}4\ M$ (6)

$$\text{Let } x = [HCO_3^-] = [H^+] = \frac{[HCO_3^-][H^+]/[H_2CO_3]}{} = x^2 / [H_2CO_3]$$
$$= 5.0e\text{-}7$$

Combining (1) and (6) gives $[H^+]^2 = x^2 = 8.0e\text{-}4 * 5.0e\text{-}7 = 4.0e\text{-}10$. Therefore,

$$[H^+] = 2.0e\text{-}5$$
$$pH = -\log(2.0e5) = 4.7$$

Discussion

Generally speaking, rain water has a pH about 5, rather acidic. It dissolves limestone and marble readily. Due to the dissolved carbon dioxide, rain water is a buffer solution.

Increased carbon dioxide level forces an increase in dissolved carbon dioxide. Would this causes pH of rain water to decrease or increase? Justify your answer by giving the reasons.

Since $[H^+] = 2.0e-5$, $[OH^-] = 5e-9$, the amount of H^+ from ionization of water is also 5.0e-9, small with respect to 2.0e-5 from ionization of H_2CO_3. Similarly, the ionization from is also small. Most of the C-containing species is H_2CO_3

$$HCO_3^- \leftrightarrow CO_3^{2-} + H^+$$

H_2CO_3 is a weak acid, its ionization is small indeed.

Now, you may proceed to evaluate other concentrations: $[OH^-]$, $[HCO_3^-]$, and $[CO_3^{2-}]$

Reactivity of Water Towards Metals

Alkali metals react with water readily. Contact of cesium metal with water causes immediate explosion, and the reactions become slower for potassium, sodium and lithium. Reaction with barium, strontium, calcium are less well known, but they do react readily. Warm water may be needed to react with calcium metal, however.

Many metals displace H^+ ions in acidic solutions. This is often seen as a property of acids.

Electrolysis of Water

The *enthalpy of formation* for liquid water, $H_2O(l)$, is -285.830 and that of water vapour is -241.826 kJ/mol. The difference is the heat of vaporization at 298 K. Liquid water and vapor entropies (S) are 69.95 and 188.835 kJ K^{-1} mol^{-1} respectively. These are entropies, not standard entropies of formation. The entropy of formation for water is obtained by,

$$\Delta S^o_{f\,water} = S^o_{water} - S^o_{H2} - 0.5\,S^o_{O2}$$

$$= 69.95 - 130.68 - 0.5*205.14$$

$$= -163.3 \text{ J K}^{-1} \text{ mol}^{-1}$$

$$\Delta G^o_{water} = \Delta H - T\Delta S \quad \text{(note } H \text{ in kJ/mol and S in J/mol)}$$

$$\Delta DG^o_{water} = -285.83 - 298.15 * 163.3/1000 = -237.13 \text{ kJ}$$

The equilibrium constant and Gibb's energy are related,

$$\Delta G^o = -R\,T \ln K$$

$$K = \exp(-DG^o / R\,T)$$

$$= 3.5e41 \text{ atm}^{-3/2}$$

This is a very large value for the formation of water,

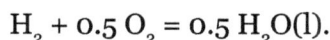

$$H_2 + 0.5\,O_2 = 0.5\,H_2O(l).$$

In other words, the reaction is complete, and the possibility of water dissociated into hydrogen and oxygen is very small. A negative value for DG^o indicates an exothermic reaction.

The Gibb's energy is the energy released other than pressure-volume work. This redox reaction to form water can be engineered to proceed in a Daniel cell. In this case, the energy is converted into electric energy according to this equation.

$$\Delta G^o_{water} = -n\,F\,E = -237.13\ kJ$$

where n is the number of electrons ($= 2$) in the redox equation, F is the Faraday constant ($= 96485$ C), and E is the potential of the Daniel cell. Thus,

$$E = -\,\frac{-\,237130\ J}{2*96485\ C}$$

$$= 1.23\ V$$

Ideally, a reverse voltage of 1.23 V is required for the electrolysis of water. But in reality, a little over voltage is required to carry out the electrolysis to decompose water. Furthermore, pure water does not conduct electricity, and acid, base or salt is often added for the electrolysis of water. This link has a demonstration.

Example 2

In order to carry out the electrolysis of water, 1.50 V is applied. Assume the energy not converted to chemical energy is converted to heat. How much heat is generated for the electrolysis of 1 mole water?

Solution

Ideally, 1.23 V will be used for the electrolysis. Energy due to the over voltage of 1.50 - 1.23 = 0.27 V is converted to heat.

$$Heat = 0.27\ V * 2 * 96485\ C$$
$$= 52102\ J$$
$$= 52\ kJ$$

Discussion

The excess energy can also be evaluated using

$$Heat = n\,F*1.50 - 237130$$

This problem also illustrates the principle of conservation of energy.

Analysis of Water Chemistry

Water chemistry analyses are carried out to identify and quantify the chemical components and properties of water samples. The type and sensitivity of the analysis depends on the purpose of the analysis and the anticipated use of the water. Chemical water analysis is carried out on water used in industrial processes, on waste-water stream, on rivers and stream, on rainfall and on the sea. In all cases the results of the analysis provides information that can be used to make decisions or to provide re-assurance that conditions are as expected. The analytical parameters selected are chosen to be appropriate for the decision making process or to establish acceptable normality. Water chemistry analysis is often the groundwork of studies of water quality, pollution, hydrology and geothermal waters. Analytical methods routinely used can detect and measure all the natural elements and their inorganic compounds and a very wide range of organic chemical species using methods such as gas chromatography and mass spectrometry. In water treatment plants producing drinking water and in some industrial processes using products with distinctive taste and odours, specialised organoleptic methods may be used to detect smells at very low concentrations.

Environmental Water

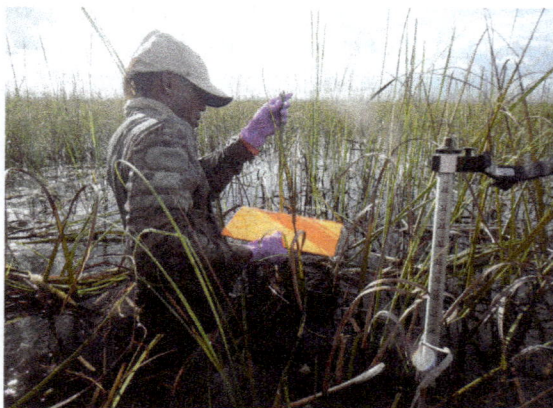

An EPA scientist samples water in Florida Everglades

Samples of water from the natural environment are routinely taken and analysed as part of a pre-determined monitoring programme by regulatory authorities to ensure that waters remain unpolluted, or if polluted, that the levels of pollution are not increasing or are falling in line with an agreed remediation plan. An example of such a scheme is the Harmonised monitoring scheme operated on all the major river systems in the UK. The parameters analysed will be highly dependent on nature of the local environment and the polluting sources in the area. In many cases the parameters will reflect the national and local water quality standards determined by law or other regulations. Typical parameters for ensuring that unpolluted surface waters remain within acceptable chemical standards include pH, major cations and anions including Ammonia, Nitrate, Nitrite, Phosphate, Conductivity, COD, Phenol and BOD.

Drinking Water Supplies

Surface or ground water abstracted for the supply of drinking water must be capable of meeting rigorous chemical standards following treatment. This requires a detailed knowledge of the water entering the treatment plant. In addition to the normal suite of environmental chemical parameters, other parameters such as hardness, Phenol, Oil and in some cases a real-time organic profile of the incoming water as in the River Dee regulation scheme.

Industrial Process Water

In industrial process, the control of the quality of process water can be critical to the quality of the end product. Water is often used as a carrier of reagents and the loss of reagent to product must be continuously monitored to ensure that correct replacement rate. Parameters measured relate specifically to the process in use and to any of the expected contaminants that may arise as by-products. This may include unwanted organic chemicals appearing in an inorganic chemical process through contamination with oils and greases from machinery. Monitoring the quality of the waste water discharged from industrial premises is a key factor in controlling and minimising pollution of the environment. In this application monitoring schemes analyse for all possible contaminants arising within the process and in addition contaminants that may have particularly adverse impacts on the environment such as Cyanide and many organic species such as pesticides. In then nuclear industry analysis focuses on specific isotopes or elements of interest. Where the nuclear industry makes waste water discharges to rivers which have drinking water abstraction on them, radio-isotopes which could potentially be harmful or those with long half-livessuch as Tritium will form part of the routine monitoring suite.

Research

Many aspects of academic research and industrial research such as in pharmaceuticals, health products, and many others relies on accurate water analysis to identify substances of potential use, to refine those substances and to ensure that when they are manufactured for sale that the chemical composition remains consistent. The analytical methods used in these area can be very complex and may be specific to the process or area of research being conducted and may involve the use of bespoke analytical equipment.

Forensic Analysis

In environmental management, water analysis is frequently deployed when pollution is suspected to identify the pollutant in order to take remedial action. The analysis can often enable the polluter to be identified. Such forensic work can examine the ratios of various components and can "type" samples of oils or other mixed organic contaminants to directly link the pollutant with the source. In drinking water supplies the cause of unacceptable quality can similarly be determined by carefully targeted chemical analysis of samples taken throughout the distribution system. In manufacturing, off-spec products may be directly tied back to unexpected changes in wet processing stages and analytical chemistry can identify which stages may be at fault and for what reason.

Methodology

To ensure consistency and repeatability, the methods use in the chemical analysis of water samples are often agreed and published at a national or state level. By convention these are often referred to as "Blue book"s.

The methods defined in the relevant standards can be broadly classified as:

- Conventional wet chemistry including Winkler method for dissolved oxygen, precipitation, filtration for solids, acidification, neutralisation titration etc. Colourimetric methods such as MBAS assay which indicates anionic surfactants in water and on site comparator methods to determine Chlorine and Chloramines. Nephelometers are used to measure solids concentrations as turbidity. These methods are generally robust and well tried and inexpensive, giving a reasonable degree of accuracy at modest sensitivity.

- Electro chemistry including pH, Conductivity and dissolved oxygen using oxygen electrode. These methods yield accurate and precise results using electronic equipment capable of feeding results directly into a laboratory data management system

- Spectrophotometry particularly for metallic elements in solution producing results with very high sensitivity but which may require some sample preparation prior to analysis and may also need specialised sampling methods to avoid sample deterioration in transit.

- Chromatography is used for many organic species which are volatile or which can yield a characteristic volatile component of after initial chemical processing.

- Ion chromatography is a sensitive and stable technique that can measure Li, NH_4 and many other low molecular weight ions using ion exchange technology.

- Gas chromatography can be used to determine methane, carbon dioxide, Cyanide oxygen and nitrogen and many other volatile components at reasonable sensitivities.

- Mass spectrometry is used where very high sensitivity is required and is sometimes used as a back-end process after Gas liquid chromatography for detecting trace organic chemicals.

Depending on the components, different methods are applied to determine the quantities or ratios of the components. While some methods can be performed with standard laboratory equipment, others require advanced devices, such as Inductively coupled plasma mass spectrometry (ICP-MS).

Metal Aquo Complexes

Metal ions in aqueous solution exist as aqua ions, where water molecules act as ligands, and coordinate to the metal ion via the oxygen donor atoms as shown for the $[Al(H_2O)_6]^{3+}$ hexaaqua ion below:

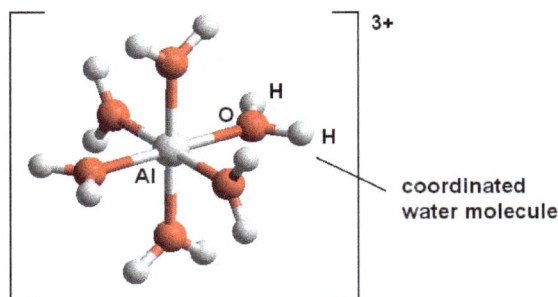

Figure: The aluminum(III) hexaaqua ion, present in aqueous solution and in many salts such as $[Al(H_2O)_6]Cl_3$, often written as $AlCl_3 \cdot 6H_2O$.

Metal ions can have varying numbers of water molecules coordinated to them, ranging from four for the very small Be(II) ion, up to 9 for the large La(III) ion. These are shown in Figure below:

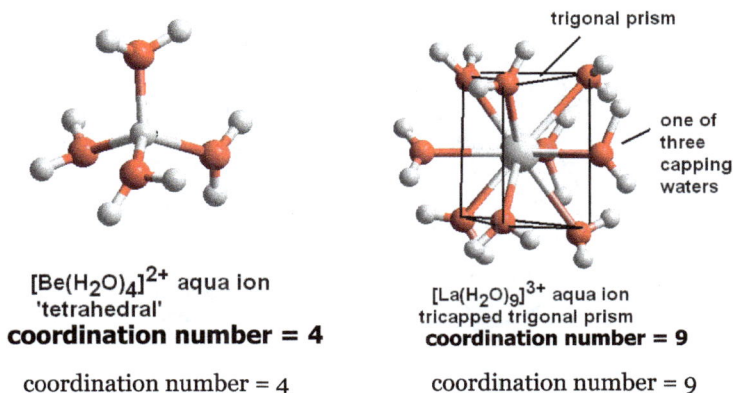

$[Be(H_2O)_4]^{2+}$ aqua ion
'tetrahedral'
coordination number = 4

coordination number = 4

$[La(H_2O)_9]^{3+}$ aqua ion
tricapped trigonal prism
coordination number = 9

coordination number = 9

Figure, The Be (II) and La(III) aqua ions, Be(II) generated using PM3. As shown, the geometry around the La^{3+} is a tricapped trigonal prism, a common geometry for nine-coordinate species with unidentate ligands.

The inner and outer sphere of waters around metal ions in solution:

In the solid state, the H-atoms of the coordinated waters are almost always H-bonded to other waters, or anions such as nitrate or perchlorate. In aqueous solution, this H-bonding structures the water molecules around the aqua ion into what is called the 'outer-sphere' of solvating water molecules, while the water molecules coordinated directly to the metal ion are referred to as the 'inner-sphere' waters. This is illustrated for the Al(III) aqua ion below, where each H-atom from an inner-sphere water has a water molecule H-bonded to it, giving twelve water molecules in the outer-sphere:

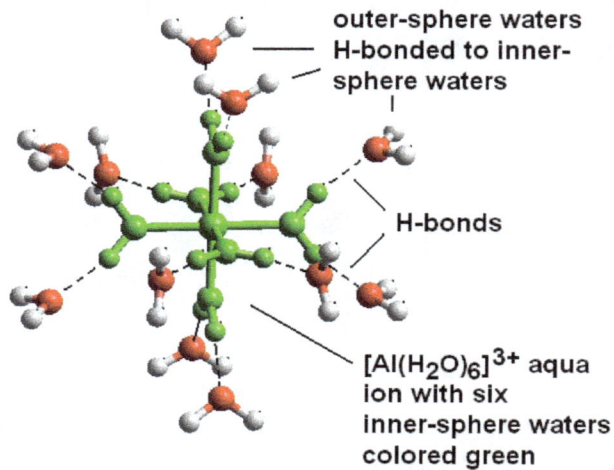

Figure: The Al(III) aqua ion showing the six inner-sphere waters (colored green) and twelve outer-sphere waters H-bonded to the inner-sphere.

Diagrammatic representation of the inner and outer sphere of waters around a metal ion in solution:

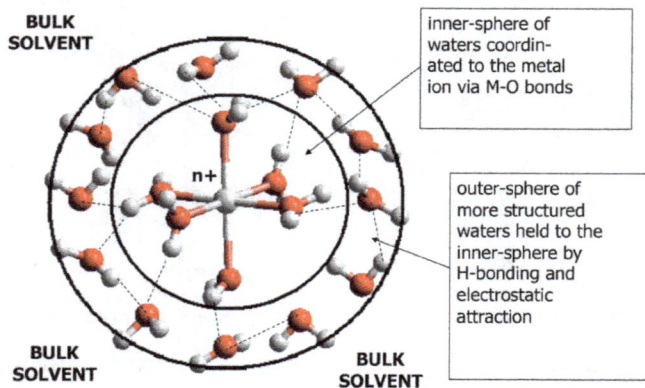

A point of interest is that water can exist also as a bridging ligand, as in numerous complexes such as those:

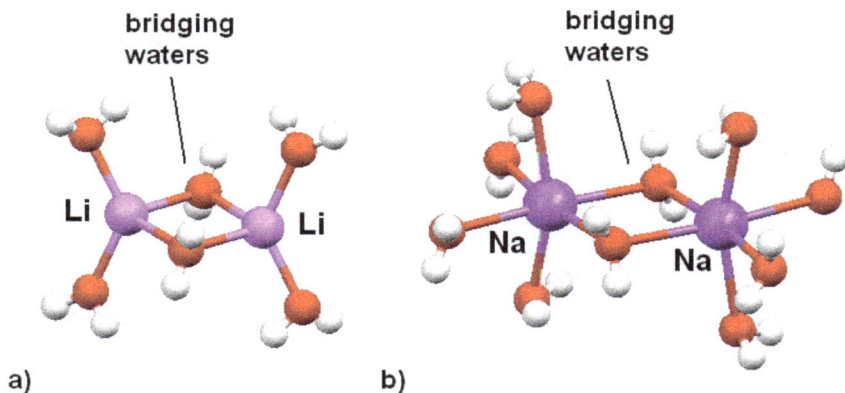

Figure: Bridging waters as found in a) the
$[Li_2(H_2O)_6]^{2+}$ cation (CSD = CELGUV) and b) the
$[Na_2(H_2O)_{10}]^{2+}$ cation (CSD = ECEPIL).

Most common are the octahedral complexes with the formula $[M(H_2O)_6]^{2+}$ and $[M(H_2O)_6]^{3+}$. Some members of this series are listed in the table below. A few aquo complexes exist with coordination numbers lower than six. Palladium(II) and platinum(II), for example, form square planar aquo complexes with the stoichiometry $[M(H_2O)_4]^{2+}$. Aquo complexes of the lanthanide trications are eight- and nine-coordinate, reflecting the large size of the metal centres.

Aquo complexes of about one third of the transition metals (Zr, Hf, Nb, Ta, Mo, W, Tc, Re, Os and Au) are either unknown or rarely described. Aquo complexes of M^{4+} centres would be extraordinarily acidic. For example, $[Ti(H_2O)_6]^{4+}$ is unknown, but $[Ti(H_2O)_6]^{3+}$ is well characterized. This acidification is related to the stoichiometry of the Zr(IV) aquo complex $[Zr_4(OH)_{12}(H_2O)_{16}]^{8+}$ (see zirconyl chloride. Similarly, $[V(H_2O)_6]^{5+}$ is unknown, but its conjugate base, $[VO(H_2O)_5]^{2+}$ is highly stable. Univalent metal centres such as Cu(I) and Rh(I) rarely form isolable complexes with water. Ag(I) forms tetrahedral $[Ag(H_2O)_4]^+$, a rare example of a tetrahedral aquo complex.

Figure: Chromium(II) ion in aqueous solution, demonstrating the pure blue color of the ion.

Some aquo complexes also contain metal-metal bonds. Two examples are $[Mo_2(H_2O)_8]^{4+}$ and $[Rh_2(H_2O)_{10}]^{4+}$.

Complex	colour	electron config.	M–O distance (Å)	water exchange rate (s^{-1}, 25 °C)	$M^{2+/3+}$ self-exchange rate ($M^{-1}s^{-1}$, 25 °C)
$[Ti(H_2O)_6]^{3+}$	violet	$(t_{2g})^1$	2.025	1.8×10^5	n.a.
$[V(H_2O)_6]^{2+}$	violet	$(t_{2g})^3$	2.12		fast
$[V(H_2O)_6]^{3+}$	yellow	$(t_{2g})^2$	1.991	5.0×10^2	fast
$[Cr(H_2O)_6]^{2+}$	blue	$(t_{2g})^3(e_g)^1$	2.06, 2.33	1.2×10^8	slow
$[Cr(H_2O)_6]^{3+}$	violet	$(t_{2g})^3$	1.961	2.4×10^{-6}	slow
$[Mn(H_2O)_6]^{2+}$	pale pink	$(t_{2g})^3(e_g)^2$	2.177	2.1×10^7	n.a.
$[Fe(H_2O)_6]^{2+}$	blue-green	$(t_{2g})^4(e_g)^2$	2.095	4.4×10^6	fast
$[Fe(H_2O)_6]^{3+}$	pale yellow	$(t_{2g})^3(e_g)^2$	1.990	1.6×10^2	fast
$[Co(H_2O)_6]^{2+}$	pink	$(t_{2g})^5(e_g)^2$	2.08	3.2×10^6	n.a.
$[Ni(H_2O)_6]^{2+}$	green	$(t_{2g})^6(e_g)^2$	2.05	3.2×10^4	n.a.
$[Cu(H_2O)_6]^{2+}$	blue	$(t_{2g})^6(e_g)^3$	1.97, 2.30	5.7×10^9	n.a.

Structure of Solid Aquo Complexes

Many transition metal salts are hexaaquo complexes, not only in solution but also as solids. Most first row divalent metals form Tutton's salts, which feature the hexaaquo complexes. One example is Ferrous ammonium sulfate.

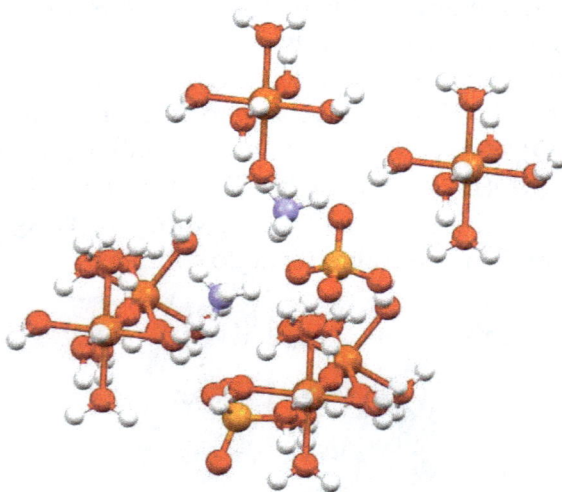

Unit cell of the Tutton salt ferrous ammonium sulfate, which features the aquo complex $[Fe(H_2O)_6)]^{2+}$ (N is violet, O is red, S is orange, Fe is large red). Related compound contain V(II), Cr(II), Mn(II), Co(II), Ni(II), and Cu(II).

Reactions

Three reactions are most fundamental to the behavior of metal aquo ions: ligand exchange, electron-transfer, and acid-base reactions of the O-H bonds.

Water Exchange

Ligand exchange involve replacement of a water ligand ("coordinated water") with water in solution ("bulk water"). Often the process is represented using labeled water H_2O^*:

$$[M(H_2O)_n]^{z+} + H_2O^* \rightarrow [M(H_2O)_{n-1}(H_2O^*)]^{z+} + H_2O$$

In the absence of isotopic labeling, the reaction is degenerate, meaning that the free energy change is zero. Rates vary over many orders of magnitude. The main factor affecting rates is charge: highly charged metal aquo cations exchange their water more slowly than singly charged species. Thus, the exchange rates for $[Na(H_2O)_6]^+$ and $[Al(H_2O)_6]^{3+}$ differ by a factor of 10^9. Electron configuration is also a major factor, illustrated by the fact that the rates of water exchange for $[Al(H_2O)_6]^{3+}$ and $[Ir(H_2O)_6]^{3+}$ differ by a factor of 10^9 also. Water exchange usually follows a dissociative substitution pathway, so the rate constants indicate first order reactions.

Electron Exchange

This reaction usually applies to the interconversion of di- and trivalent metal ions, which involves the exchange of only one electron. The process is called self-exchange, meaning that the ion *appears* to exchange electrons with itself. The standard electrode potential for the equilibrium

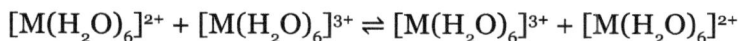

$$[M(H_2O)_6]^{2+} + [M(H_2O)_6]^{3+} \rightleftharpoons [M(H_2O)_6]^{3+} + [M(H_2O)_6]^{2+}$$

Standard redox potential for the couple M^{2+}, M^{3+} / V				
V	Cr	Mn	Fe	Co
-0.26	-0.41	+1.51	+0.77	+1.82

shows the increasing stability of the lower oxidation state as atomic number increases. The very large value for the manganese couple is a consequence of the fact that octahedral manganese(II) has zero crystal field stabilization energy (CFSE) but manganese(III) has 3 units of CFSE.

Using labels to keep track of the metals, the self-exchange process is written as:

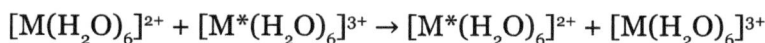

$$[M(H_2O)_6]^{2+} + [M^*(H_2O)_6]^{3+} \rightarrow [M^*(H_2O)_6]^{2+} + [M(H_2O)_6]^{3+}$$

The rates of electron exchange vary widely, the variations being attributable to differing reorganization energies: when the 2+ and 3+ ions differ widely in structure, the rates tend to be slow. The electron transfer reaction proceeds via an outer sphere electron transfer. Most often large reorganizational energies are associated with changes in the population of the e_g level, at least for octahedral complexes.

Acid–base Reactions

Solutions of metal aquo complexes are acidic owing to the ionization of protons from the water ligands. In dilute solution chromium(III) aquo complex has a pK_a of about 4.3:

$$[Cr(H_2O)_6]^{3+} \rightleftharpoons [Cr(H_2O)_5(OH)]^{2+} + H^+$$

Thus, the aquo ion is a weak acid, of comparable strength to acetic acid (pK_a of about 4.8). This is typical of the trivalent ions. The influence of the electronic configuration on acidity is shown by the fact that $[Ru(H_2O)_6]^{3+}$ ($pK_a = 2.7$) is more acidic than $[Rh(H_2O)_6]^{3+}$ ($pK_a = 4$), despite the fact that Rh(III) is expected to be more electronegative. This effect is related to the stabilization of the pi-donor hydroxide ligand by the $(t_{2g})^5$ Ru(III) centre.

In more concentrated solutions, some metal hydroxo complexes undergo condensation reactions, known as olation, to form polymeric species. Many minerals, form via olation. Aquo ions of divalent metal ions are less acidic than those of trivalent cations.

The hydrolyzed species often exhibit very different properties from the precursor hexaaquo complex. For example, water exchange in $[Al(H_2O)_5OH]^{2+}$ is some 20000 times faster than in $[Al(H_2O)_6]^{3+}$.

Hydrolysis

hydrolysis is a chemical reaction in which water is used to break down the bonds of a particular substance. In biotechnology and as far as living organisms are concerned, these substances are often polymers.

The word hydrolysis comes from the word hydro, which is Greek for water, and lysis, which means "to unbind." In practical terms, hydrolysis means the act of separating chemicals when to water is added. There are three main types of hydrolysis: salt, acid, and base hydrolysis.

Hydrolysis can also be thought of as the exact opposite reaction to condensation, which is the process whereby two molecules combine to form one larger molecule. The end result of this reaction is that the larger molecule ejects a water molecule. You will always remember the difference between the two if you think of it in the context that hydrolysis uses water to break down something while condensation, on the other hand, grows something, by removing water.

Common Types of Hydrolysis

- Salts: Hydrolysis occurs when salt from a weak base or acid dissolves in liquid.

When this occurs, water spontaneously ionizes into hydroxide anions and hydronium cations. This is the most common type of hydrolysis.

- Acid: Water can act as an acid or a base, according to the Bronsted-Lowry acid theory. In this case, the water molecule would give away a proton. Perhaps the oldest commercially-practiced example of this type of hydrolysis is saponification, the formation of soap.

- Base: This reaction is very similar to the hydrolysis for base dissociation. Again, on a practical note, a base that often dissociates in water is ammonia.

Hydrolysis Reaction

In a hydrolysis reaction involving an ester link, such as that found between two amino acids in a protein, the products that result include one that receives the hydroxyl (OH) group from the water molecule and another that becomes a carboxylic acid with the addition of the remaining proton (H^+).

Hydrolysis Reactions in Living Organisms

Hydrolysis reactions in living organisms are performed with the help of catalysis by a class of enzymes known as hydrolases. The biochemical reactions that break down polymers, such as proteins (which are peptide bonds between amino acids), nucleotides, complex sugars and starch, and fats are catalyzed by this class of enzymes. Within this class are lipases, amylases, proteinases hydrolyze fats, sugars, and proteins, respectively.

Cellulose-degrading bacteria and fungi play a special role in paper production and other everyday biotechnology applications because they have enzymes (such as cellulases and esterases) that can break cellulose into polysaccharides (i.e., polymers of sugar molecules) or glucose, and break down stickies.

For example, Proteinase was added to the cell extract, in order to hydrolyze the peptides and produce a mixture of free amino acids.

Salt Hydrolysis

Equations

A salt is an ionic compound that is formed when an acid and a base neutralize each other. While it may seem that salt solutions would always be neutral, they can frequently be either acidic or basic.

Consider the salt formed when the weak acid hydrofluoric acid is neutralized by the strong base sodium hydroxide. The molecular and net ionic equations are shown below.

$$HF(aq) + NaOH(aq) \rightarrow NaF(aq) + H_2O(l)$$
$$HF(aq) + OH^-(aq) \rightarrow F^-(aq) + H_2O(l)$$

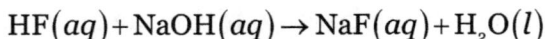

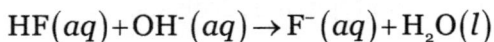

Since sodium fluoride is soluble, the sodium ion is a spectator ion in the neutralization reaction. The fluoride ion is capable of reacting, to a small extent, with water, accepting a proton.

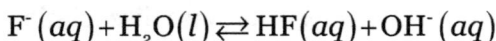

$$F^-(aq) + H_2O(l) \rightleftharpoons HF(aq) + OH^-(aq)$$

The fluoride ion is acting as a weak Brønsted-Lowry base. The hydroxide ion that is produced as a result of the above reaction makes the solution slightly basic. Salt hydrolysis is a reaction in which one of the ions from a salt reacts with water, forming either an acidic or basic solution.

Salts that form Basic Solutions

When solid sodium fluoride is dissolved into water, it completely dissociates into sodium ions and fluoride ions. The sodium ions do not have any capability of hydrolyzing, but the fluoride ions hydrolyze to produce a small amount of hydrofluoric acid and hydroxide ion.

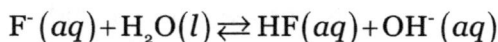

$$F^-(aq) + H_2O(l) \rightleftharpoons HF(aq) + OH^-(aq)$$

Salts that are derived from the neutralization of a weak acid (HF) by a strong base (NaOH) will always produce salt solutions that are basic.

Salts that form Acidic Solutions

Ammonium chloride (NH_4Cl) is a salt that is formed when the strong acid HCl is neutralized by the weak base NH_3. Ammonium chloride is soluble in water. The chloride ion produced is incapable of hydrolyzing because it is the conjugate base of the strong acid HCl. In other words, the Cl^- ion cannot accept a proton from water to form HCl and OH^-, as the fluoride ion. However, the ammonium ion is capable of reacting slightly with water, donating a proton and so acting as an acid.

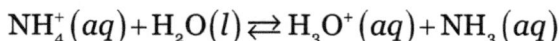

$$NH_4^+(aq) + H_2O(l) \rightleftharpoons H_3O^+(aq) + NH_3(aq)$$

Salts that form Neutral Solutions

A salt that is derived from the reaction of a strong acid with a strong base forms a solution that has a pH of 7. An example is sodium chloride, formed from the neutralization of HCl by NaOH. A solution of NaCl in water has no acidic or basic properties, since neither ion is capable of hydrolyzing. Other salts that form neutral solutions include potassium nitrate (KNO$_3$) and lithium bromide (LiBr). The Table below summarizes how to determine the acidity or basicity of a salt solution.

Salt formed from:	Salt Solution
Strong acid + Strong base	Neutral
Strong acid + Weak base	Acidic
Weak acid + Strong base	Basic

Salts formed from the reaction of a weak acid and a weak base are more difficult to analyze because of competing hydrolysis reactions between the cation and the anion.

Hydrolysis in Acid Reactions

It can get a little confusing to think about which reactions are acid reactions and which are basic reactions because for acidic hydrolysis the water is acting as a base, while for basic hydrolysis the water is acting as an acid. The way to keep this straight is to remember that hydrolysis refers to a molecule being broken with water. So what molecule is being broken? In acidic hydrolysis, it is an acidic molecule that is being broken. For the acidic molecule to be broken, the water (in the hydrolysis portion of the name) needs to act as a base. It works in the opposite way for basic hydrolysis.

Let's look at some acidic hydrolysis examples. For acidic hydrolysis to work, we react water with a weak acid such as acetic acid or hydrofluoric acid.

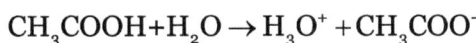

$$CH_3COOH + H_2O \rightarrow H_3O^+ + CH_3COO^-$$

The acetic acid reacts with the water, and the water acts as a base and accepts the hydrogen, breaking the oxygen hydrogen bond.

$$HF + H_2O \rightarrow F^- + H_3O^+$$

In this example, the hydrofluoric acid reacts with the water, and the water acts as a base and accepts the hydrogen from the hydrofluoric acid. This breaks the fluoride-hydrogen bond, and we end up with a fluoride ion and a hydronium ion.

These reactions can now act as a catalyst for other reactions since we have a positive charge on the water and a negative charge on the acetic acid. Both of these compounds can now easily react with other compounds. For example, the water (hydronium ion) can react with an ester and break it into a carboxylic acid and an alcohol.

This hydronium ion is also what reacts with ATP to break the phosphate bond to release energy into the body.

Hydrolysis and Basic Salts

A basic salt is formed between a weak acid and a strong base. The basicity is due to the hydrolysis of the conjugate base of the (weak) acid used in the neutralization reaction. For example, sodium acetate formed between the weak acetic acid and the strong base NaOH is a basic salt. When the salt is dissolved, ionization takes place:

$$NaAc = Na^+ + Ac^-$$

In the presence of water, Ac^- undergo hydrolysis:

$$H_2O + Ac^- = HAc + OH^-$$

And the equilibrium constant for this reaction is K_b of the conjugate base Ac^- of the acid HAc. Note the following equilibrium constants:

$$K_b = \frac{[HAc]\,[OH^-]}{[Ac^-]}$$

$$K_b = \frac{[HAc]\,[OH^-]\,[H^+]}{[Ac^-]\quad[H^+]}$$

$$K_b = \frac{[HAc]}{[Ac^-]}\,\frac{[OH^-]\,[H^+]}{[H^+]}$$

$$= K_w / K_a$$

$$= 1.00\text{e-}14 / 1.75\text{e-}5 = 5.7\text{e-}10.$$

Thus,

$$K_a K_b = K_w$$

or

$$pK_a + pK_b = 14$$

for a conjugate acid-base pair. Let us look at a numerical problem of this type.

Example

Calculate the $[Na^+]$, $[Ac^-]$, $[H^+]$ and $[OH^-]$ of a solution of 0.100 M NaAc (at 298 K). (K_a = 1.8E-5)

Solution

Let x represent $[H^+]$, then

$$H_2O + Ac^- = HAc + OH^-$$

0.100-x x   x

$$\frac{x^2}{0.100-x} = (1E-14)/(1.8E-5) = 5.6E-10$$

Solving for x results in

x = sqrt(0.100*5.6E-10)

  = 7.5E-6

$[OH^-] = [HAc] = 7.5E-6$

$[Na^+] = 0.100$ F

Discussion

This corresponds to a pH of 8.9 or $[H^+] = 1.3E-$.

Note that $K_w / K_a = K_b$ of Ac^-, so that K_b rather than K_a may be given as data in this question.

Water of Crystallization

Water of crystallization is the fixed amount of water, which is necessary for certain salts to crystallize out from their aqueous solutions. This water makes it possible for them to form crystals, and it is responsible for the shapes of their crystals.

The crystals contain the salts and water combined in definite proportions, and when heated, the water is lost. Example, $CuSO_4.5H_2O$ (blue crystals) - the crystals are formed by the chemical combination of 1 mole of $CuSO_4$ with 5 moles of H_2O.

When heated, it looses the water of crystallization (i.e. 5 moles) and forms a white powder, $CuSO_4$. Other examples of salts which crystallize with water are $Na_2CO_3.10H_2O$ (crystallizes with 10 moles of H_2O); $Na_2SO_4.10H_2O$ (with 10 moles of H_2O) and $FeSO_4.7H_2O$ (with 7 moles of H_2O).

From studies, it is found that the water of crystallization actually form chemical bonds with the positive and negative ions of the salts. Example, in $CuSO_4.5H_2O$, 4 moles of H_2O are in coordinate bonding with Cu^{2+}, while 1 mole of water forms hydrogen bonding with the SO_4^{2-} ion. In some hydrated salts however, some of the water do not get attached by chemical bonds, but occupy certain positions in the crystal structure - such water is called lattice water.

An example is the hydrate $KAl(SO_4)_2.12H_2O$, in which $6H_2O$ molecules are lattice water molecules and the other 6 are water of crystallization because they are attached by co-ordinate bonds to the aluminium ion. Example: If 0.715 g of a hydrated form of sodium trioxocarbonate exactly reacts with 50 cm³ of 0.10 M hydrochloric acid, determine the number of moles of water of crystallization present in one mole of the hydrated salt. (Na = 23, C = 12, O = 16, H = 1)

Solution: Take note of the following:

1. The mass of the hydrated salt 0.715 g is the sum of the masses of the anhydrous salt and the water of crystallization present.

2. Only the anhydrous salt will react with the hydrochloric acid. The water molecules present as water of crystallization do not get involved in the reaction. Number of moles of HCl involved in the reaction:

M = No. of moles/V(dm³)

No. of moles = 0.10 x 0.05 dm³ = 0.005 mole

Therefore, from the equation of the reaction:

$Na_2CO_3(s) + 2HCl(aq) → 2NaCl(aq) + H_2O(l) + CO_2(g)$

1 mole of Na_2CO_3 reacted with 2 moles of HCl, therefore, 0.0025 (i.e. 0.005/2) mole of Na_2CO_3 reacted with 0.005 mole of HCl.

Converting 0.0025 mole of Na_2CO_3 to mass in grams, we have:

Mass (g) = No. of moles x molar mass = 0.0025 x 106 = 0.265 g

Hence, mass of water of crystallization = 0.715 - 0.265 = 0.45 g

No. of moles of water of crystallization = 0.45/18 = 0.025 mole

Therefore, the mole ratio between Na_2CO_3 and H_2O = 0.0025 : 0.025

I.e. 0.0025/0.0025 : 0.025/0.0025

1 : 10

Therefore, the molecular formula of the hydrated salt is $Na_2CO_3 . 10H_2O$

Hence, in one mole of the hydrated salt 10 moles of water of crystallization are contained.

Note: The combination between the water of crystallization and the salt ions is a chemical reaction - a proof is that certain amount of heat (heat of hydration) is given off when a salt crystallizes from its solution.

Also, a change is observed in both the appearance and the mass of the salt, before and after crystallization (i.e., the anhydrous or powdery salt compared with the crystalline form.

There is no difference between the solutions obtained by dissolving both anhydrous salt and salt with water of crystallization. Example, the solution of $CuSO_4$ is same as that of $CuSO_4 \cdot 5H_2O$.

Hydration Reaction

In chemistry, a hydration reaction is a chemical reaction in which a substance combines with water. In organic chemistry, water is added to an unsaturated substrate, which is usually an alkene or an alkyne. This type of reaction is employed industrially to produce ethanol, isopropanol, and 2-butanol.

Epoxides to Glycol

Several billion kilograms of ethylene glycol are produced annually by the hydration of oxirane, a cyclic compound also known as ethylene oxide:

$$C_2H_4O + H_2O \rightarrow HO-CH_2CH_2-OH$$

Acid catalysts are typically used.

Alkenes

For the hydration of alkenes, the general chemical equation of the reaction is the following:

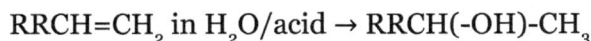

$$RRCH=CH_2 \text{ in } H_2O/acid \rightarrow RRCH(-OH)-CH_3$$

A hydroxyl group (OH^-) attaches to one carbon of the double bond, and a proton (H^+) adds to the other carbon of the double bond. The reaction is highly exothermic. In the first step, the alkene acts as a nucleophile and attacks the proton, following Markovnikov's rule. In the second step an H_2O molecule bonds to the other, more highly substituted carbon. The oxygen atom at this point has three bonds and carries a positive charge (i.e., the molecule is an oxonium). Another water molecule comes along and takes up the extra proton. This reaction tends to yield many undesirable side products, (for example diethyl ether in the process of creating Ethanol) and in its simple form described here is not considered very useful for the production of alcohol.

Two approaches are taken. Traditionally the alkene is treated with sulfuric acid to give alkyl sulfate esters. In the case of ethanol production, this step can be written:

$$H_2SO_4 + C_2H_4 \rightarrow C_2H_5\text{-O-SO}_3H$$

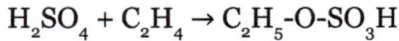

Subsequently, this sulfate ester is hydrolyzed to regenerate sulfuric acid and release ethanol:

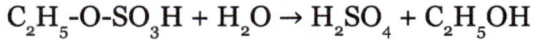

$$C_2H_5\text{-O-SO}_3H + H_2O \rightarrow H_2SO_4 + C_2H_5OH$$

This two step route is called the "indirect process".

In the "direct process," the acid protonates the alkene, and water reacts with this incipient carbocation to give the alcohol. The direct process is more popular because it is simpler. The acid catalysts include phosphoric acid and several solid acids. Here an example reaction mechanism of the hydration of 1-methylcyclohexene to 1-methylcyclohexanol:

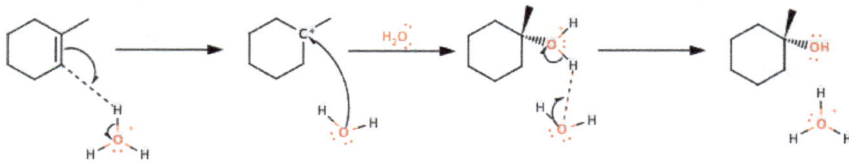

Many alternative routes are available for producing alcohols, including the hydroboration–oxidation reaction, the oxymercuration–reduction reaction, the Mukaiyama hydration, the reduction of ketones and aldehydes and as a biological method fermentation.

Hydration of other Substrates

Any unsaturated organic compound is susceptible to hydration. Acetylene hydrates to give acetaldehyde: The process typically relies on mercury catalysts and has been discontinued in the West but is still practiced in China. The Hg^{2+} center binds to C≡C triple bond, which is then attacked by water. The reaction is:

$$H_2O + C_2H_2 \rightarrow CH_3CHO$$

Nitriles undergo hydration to give amides:

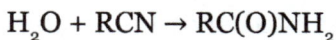

$$H_2O + RCN \rightarrow RC(O)NH_2$$

This reaction is employed in the production of acrylamide.

Aldehydes and to some extent even ketones, hydrate to geminal diols. The reaction is especially dominant for formaldehyde, which, in the presence of water, exists significantly as dihydroxymethane.

Conceptually similar reactions include hydroamination and hydroalkoxylation, which involve adding amines and alcohols to alkenes.

Dehydration Reaction

In chemistry and the biological sciences, a dehydration reaction, also known as Zimmer's hydrogenesis, is a chemical reaction that involves the loss of a water molecule from the reacting molecule. Dehydration reactions are a subset of condensation reactions. Because the hydroxyl group (−OH) is a poor leaving group, having a Brønsted acid catalyst often helps by protonating the hydroxyl group to give the better leaving group, $-OH_2^+$. The reverse of a dehydration reaction is a hydration reaction. Common dehydrating agents used in organic synthesis include concentrated sulfuric acid, concentrated phosphoric acid, hot aluminium oxide and hot ceramic.

Dehydration reactions and dehydration synthesis have the same meaning, and are often used interchangeably. Two monosaccharides, such as glucose and fructose, can be joined together using dehydration synthesis. The new molecule, consisting of two monosaccharides, is called a disaccharide.

The process of hydrolysis is the reverse reaction, meaning that the water is recombined with the two hydroxyl groups and the disaccharide reverts to being monosaccharides.

In the related condensation reaction water is released from two different reactants.

Dehydration Reactions

In organic synthesis, there are many examples of dehydration reaction, for example dehydration of alcohols or sugars.

Reaction	Equation	
Conversion of alcohols to ethers	$2\ R-OH \rightarrow R-O-R + H_2O$	
Fischer–Speier esterification	$R-COOH + R'-OH \rightarrow R-COO-R' + H_2O$	
Conversion of alcohols to alkenes	$R-CH_2-CHOH-R \rightarrow R-CH=CH-R + H_2O$	for example the conversion of glycerol to acrolein: or the dehydration of 2-methyl-1-cyclohexanol to (mainly) 1-methylcyclohexene

Conversion of carboxylic acids to acid anhydrides	$2RCOOH \rightarrow (RCO)_2O + H_2O$	
Conversion of amides to nitriles	$RCONH_2 \rightarrow R-CN + H_2O$	
Dienol benzene rearrangement		

Some dehydration reactions can be mechanistically complex, for instance the reaction of a sugar (sucrose) with concentrated sulfuric acid: to form carbon as a graphitic foam involves formation of carbon-carbon bonds. The reaction is driven by the strongly exothermic reaction as sulfuric acid reacts with water, which produces dangerous sulfuric-acid containing steam, therefore the experiment should only be performed in a fume-hood or well ventilated area.

Other examples of dehydration synthesis reactions are the formation of triglycerides from fatty acids and the formation of glycosidic bonds between carbohydrate molecules, such as the formation of maltose from two glucose molecules.

Heterogenous Water Oxidation

Heterogeneous Water Oxidation

Water oxidation is one of the half reactions of water splitting:

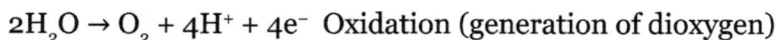

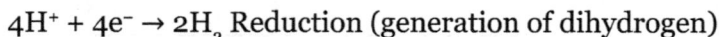

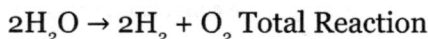

$$2H_2O \rightarrow O_2 + 4H^+ + 4e^- \text{ Oxidation (generation of dioxygen)}$$

$$4H^+ + 4e^- \rightarrow 2H_2 \text{ Reduction (generation of dihydrogen)}$$

$$2H_2O \rightarrow 2H_2 + O_2 \text{ Total Reaction}$$

Of the two half reactions, the oxidation step is the most demanding because it requires the coupling of 4 electron and proton transfers and the formation of an oxygen-oxygen bond. This process occurs naturally in plants photosystem II to provide protons and electrons for the photosynthesis process and release oxygen to the atmosphere. Since hydrogen can be used as an alternative clean burning fuel, there has been a need to split water efficiently. However, there are known materials that can mediate the reduction step efficiently therefore much of the current research is aimed at the oxidation half reaction also known as the Oxygen Evolution Reaction (OER). Current research focuses on understanding the mechanism of OER and development of new materials that catalyze the process.

Thermodynamics

Both the oxidation and reduction steps are pH dependent.

2 half reactions (at pH = 0)

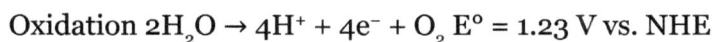

Oxidation $2H_2O \rightarrow 4H^+ + 4e^- + O_2$ $E° = 1.23$ V vs. NHE

Reduction $4H^+ + 4e^- \rightarrow 2H_2$ $E° = 0.00$ V vs. NHE

Overall $2H_2O \rightarrow 2H_2 + O_2$ $E°cell = 1.23$ V; $\Delta G = -475$ kJ/mol

Water splitting can be done at higher pH values as well however the standard potentials will vary according to the Nernst equation and therefore shift by -59 mV for each pH unit increase. However, the total cell potential (difference between oxidation and reduction half cell potentials) will remain 1.23 V. This potential can be related to Gibbs free energy (ΔG) by:

$$\Delta G°cell = -nFE°cell$$

Where n is the number of electrons per mole products and F is the Faraday constant. Therefore, it takes 475 kJ of energy to make one mole of O2 as calculated by thermodynamics. However, in reality no process can be this efficient. Systems always suffer from an overpotential that arise from activation barriers, concentration effects and voltage drops due to resistance. The activation barriers or activation energy is associated with high energy transition states that are reached during the electrochemical process of OER. The lowering of these barriers would allow for OER to occur at lower overpotentials and faster rates.

Mechanism

Heterogeneous OER is sensitive to the surface which the reaction takes place and is also affected by the pH of the solution. The general mechanism for acidic and alkaline solutions is shown below. Under acidic conditions water binds to the surface with the irreversible removal of one electron and one proton to form a platinum hydroxide. In an alkaline solution a reversible binding of hydroxide ion coupled to a one electron oxidation is thought to precede a turnover-limiting electrochemical step involving the removal of one proton and one electron to form a surface oxide species. The shift in mechanism between the pH extremes has been attributed to the kinetic facility of oxidizing hydroxide ion relative to water. Using the Tafel equation, one can obtain kinetic information about the kinetics of the electrode material such as the exchange current density and the Tafel slope. OER is presumed to not take place on clean metal surfaces such as Platinum, but instead an oxide surface is formed prior to oxygen evolution.

OER under acidic conditions.

OER under alkaline conditions.

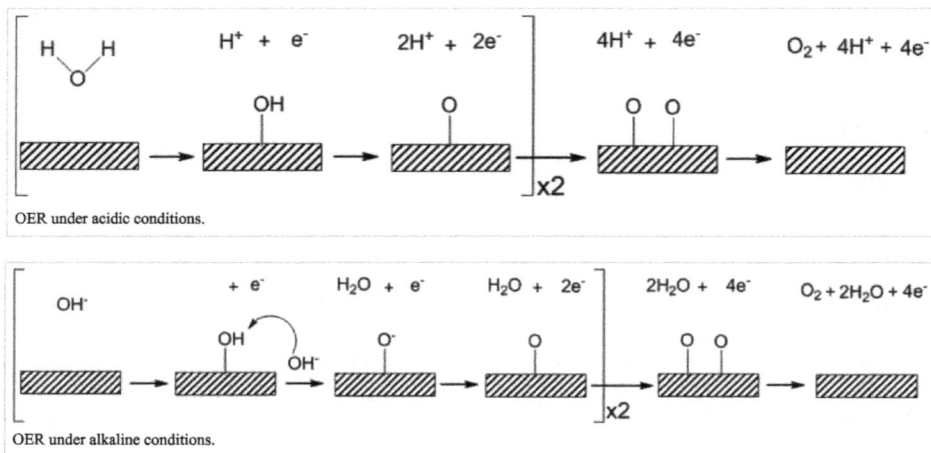

Catalyst Materials

OER has been studied on a variety of materials including:

- Platinum surfaces

- Transition metal oxides

- First-row transition metal spinels and perovskites. Recently Metal-Organic Framework (MOF)-based materials have been shown to be a highly promising candidate for water oxidation with first row transition metals.;

Preparation of the surface and electrolysis conditions have a large effect on reactivity (defects, steps, kinks, low coordinate sites) therefore it is difficult to predict an OER material's properties by its bulk structure. Surface effects have a large influence on the kinetics and thermodynamics of OER.

Platinum

Platinum has been a widely studied material for OER because it is the catalytically most active element for this reaction. It exhibits exchange current density values on the order of 10^{-9} A/cm^2. Much of the mechanistic knowledge of OER was gathered from studies on platinum and its oxides. It was observed that there was a lag in the evolution of oxygen during electrolysis. Therefore an oxide film must first form at the surface before OER begins. The Tafel slope, which is related to the kinetics of the electrocatalytic reaction, was shown to be independent of the oxide layer thickness at low current densities but becomes dependant on oxide thickness at high current densities.

Ruthenium Oxide

Ruthenium oxide (RuO_2) shows some of the best performance as an OER material in acidic environments. It has been studied since the early 1970s as a water oxidation catalyst with one of the lowest reported overpotentials for OER at the time. It has since been investigated for OER in Ru(110) single crystal oxide surfaces, compact films, Titanium supported films. RuO_2 films can be prepared by thermal decomposition of ruthenium chloride on inert substrates.

Electrolysis of Water

Creating an electric potential through water causes positive ions, including the inherent hydrogen ions H_3O^+, to move towards the negative electrode (cathode) and negative ions, including the inherent hydroxide ions OH^-, to move towards the positive electrode (anode). With a sufficient potential difference, this may cause electrolysis with oxygen gas being produced at the anode and hydrogen gas produced at the cathode. The electrolysis of water usually involves dilute, or moderately concentrated, salt solutions in order to reduce the power loss driving the current through the solution, but the presence of salt is not a requirement for electrolysis.

Thus,

Anode	+ve	$6H_2O(l) \rightarrow O_2(g) + 4H_3O^+(aq) + 4e^-(to\ anode)$ [b]	$E° = +1.229$ V, pH 0 [d]	$E°_> = +0.815$ V
Cathode	-ve	$4e^-(from\ cathode) + 4H_2O(l) \rightarrow 2H_2(g) + 4OH^-$ (aq)	$E° = -0.828$ V, pH 14	$E°_> = -0.414$ V
Overall		$2H_2O(l) \rightarrow 2H_2(g) + O_2(g)$	$\Delta G°' = +474.3$ kJ $\times$ mol^{-1}	

Where (l), (g) and (aq) show the states of the material as being a liquid, a gas or an aqueous solution and with the electrical circuit passing the electrons back from the anode to the cathode. The reactions are heterogeneous, taking place at the boundary between the electrode and the electrolyte with the aqueous boundary layer subject to concentration and electrical potential gradients with the presence of the generated gaseous nanobubbles and microbubbles.

Water Electrolysis Electrode Potentials with pH

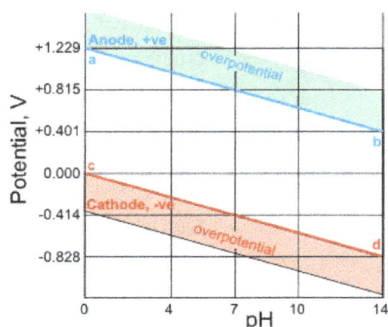

Generally, the water next to the electrodes c will change pH due to the ions produced or consumed. If the electrode compartments are separated by a suitable porous membrane then the concentration of H_3O^+ in the anolyte and OH^- in the catholyte (and hence the increase in the respective conductivities) are both expected to increase more than if there is free mixing between the electrodes, when most of these ions will neutralize each other. Small but expected differences in the solutions' pHs next to the anode (anolyte) and cathode (catholyte) cause only a slight change to the overall potential difference required (1.229 V). Increasing the acid content next to the anode due to the H_3O^+ produced will increase its electrode potential (for example: pH 4 E = +0.992 V) and increasing the alkaline content next to the cathode due to the OH^- produced will make its electrode potential more negative (for example: pH 10 E = -0.592 V). If the anode reaction is forced to run at pH 14 and the cathode reaction is run at pH 0.0, then the electrode potentials are +0.401 V and 0 V respectively.

(a) Anode	pH 0	$2 H_2O \rightarrow O_2 + 4H^+ + 4e^-$	$E° = +1.229$ V
(b) Anode	pH 14	$4 OH^- \rightarrow O_2 + H_2O + 4e^-$	$E° = +0.401$ V
(c) Cathode	pH 0	$4 H^+ + 4e^- \rightarrow 2H_2$	$E° = 0.0$ V
(d) Cathode	pH 14	$4 H_2O + 4e^- \rightarrow 2H_2 + 4OH^-$	$E° = -0.828$ V

This does not mean that because the electrolysis can be achieved with a (minimum) voltage of +0.403 V, it breaks the thermodynamic requirement of 1.229 V as there is a further input of energy required in keeping the electrode compartments at the required pHs and solute concentration.

The current flowing indicates the rate of electrolysis. The amount of product formed can be calculated directly from the duration and current flowing, as 96,485 coulombs (*i.e.*, one faraday) delivers one mole of electrons; with one faraday ideally producing 0.5 moles of H_2 plus 0.25 moles of O_2. Thus, one amp flowing for one second (one coulomb) produces 5.18 µmol H_2 (10.455 µg, 0.1177 mL at STP) and 2.59 µmol O_2 (82.888 µg, 0.0588 mL at STD; 4.9 kW h/m³ H_2 at 60% efficiency), if there are no side reactions at the electrodes; that is

Number of moles = Coulombs/(unsigned numeric charge on the ion × Faraday)
Number of moles = (Current in amperes × time in seconds)/(unsigned numeric charge on the ion × Faraday)

The gases produced at the electrodes may dissolve, with their equilibrium solubility proportional to their partial pressure as gases in the atmosphere above the electrolytic surface. Oxygen gas is poorly soluble ($\approx$ 44 mg kg⁻¹, $\approx$ 1.4 mM at 0.1 MPa and 20 °C, but only $\approx$ 0.29 mM against its normal atmospheric partial pressure). Hydrogen gas is less soluble ($\approx$ 1.6 mg kg⁻¹, $\approx$ 0.80 mM at 0.1 MPa and 20 °C but only $\approx$ 0.44 nM against its very low normal atmospheric partial pressure). It may take considerable time for the solubilities to drop from their initially-super-saturated state to their equilibrium values after the electrolysis has ended.

Although theoretically as described above, the current passing should determine the amounts of hydrogen and oxygen formed, several factors ensure that somewhat lower amounts of gas are actually found:

(i) Some electrons (and products) are used up in side-reactions,

(ii) Some of the products are catalytically reconverted to water at the electrodes, particularly if there is no membrane dividing the electrolysis compartments,

(iii) Some hydrogen may absorb into the cathode (particularly if palladium is used),

(iv) Some oxygen oxidizes the anode,

(v) Some gas remains held up in the nanobubbles for a considerable time, and

(vi) Some gas may escape measurement.

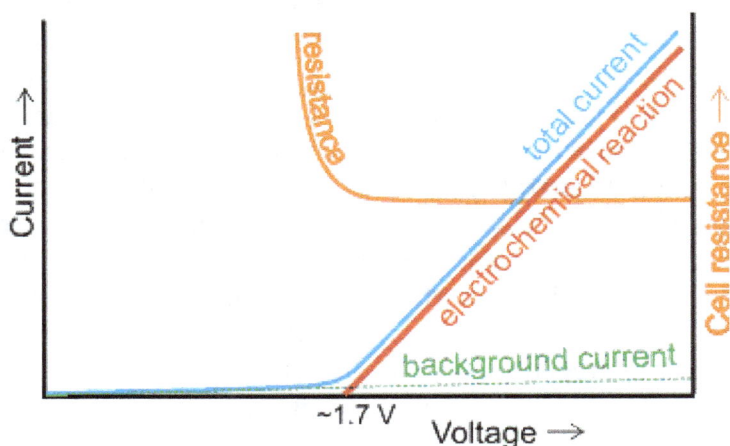

The above description hides much important science and grossly over-simplifies the system. The actual potential required at any position within the electrolytic cell is determined by the localized concentration of the reactants and products including the local pH of the solution, instantaneous gas partial pressure and effective electrode surface area loss due to attached gas bubbles. In addition, a greater potential difference is required at both electrodes to overcome the activation energy barriers and insulating bubble coverage, and then to deliver a significant reaction rate. Typically at good electrodes, such as those made of platinum, that may total an addition of about half a volt to the potential difference between the electrodes. In addition, a further potential difference is required to drive the current through the electrical resistance of electrolytic cell and circuit; for a one-ohm cell circuit resistance, a each amp current flow would require a further one volt and waste one watt of power. This power loss causes the electrolyte to warm up during electrolysis.

To clarify:

> The minimum necessary cell voltage to start water electrolysis is the potential 1.229 V.
>
> The potential necessary to start water electrolysis without withdrawing heat from the surroundings is
>
> $$-\Delta H° / nF = 1.481v$$
>
> This results in at least a 21% unavoidable loss of efficiency. Normally further heat is generated, and efficiency lost, from the over potentials applied. Additionally, energy is lost due to the evaporation of water within the wet gasses evolved.

The efficiency of electrolysis increases with the temperature as the hydrogen bonding reduces. However, the heat demand increases, due to the endergonic process, as the electrical demand decreases; mostly balancing in terms of overall energy demand. If the pressure over the electrolysis is increased then more current passes for the same applied voltage. However, the output of gas per coulomb and the heating effect are both decreased. This is due to the increased solubility of the gases and smaller bubbles both reducing the cell resistance and increasing recombination reactions. Although reducing the distance between electrodes reduces the resistance of the electrolysis medium, the process may suffer if the closeness allows a build-up of gas between these electrodes. Low to higher pulsed potential increases the reaction (current) and accelerates both the movement of bubbles from the electrode surface and the mass transfer rate in the electrolyte, which lowers the electrochemical polarization in the diffusion layer and further increases hydrogen production efficiency. The rate of change of the current density (and hence efficiency) can be increased using a magnetic field with or without optical enhancement.

Pure water conducts an electric current very poorly and, for this reason, is difficult (slow) to electrolyze. Usually, however, some salts will be added or present in tap and ground waters which will be sufficient to allow electrolysis to proceed at a significant rate. However such salts, and particularly chloride ions, may then undergo redox reactions at an electrode. These side reactions both reduce the efficiency of the electrolysis reactions (above) and produce new solutes. Other electrolytic reactions may occur at the electrodes so producing further solutes and gases. In addition, these solutes may react together to produce other materials. Together the side reactions are complex and this complexity increases somewhat when the voltage applied to the cell is greater than that required by the above reactions and processes. The likely reactions within the electrode compartments are described below. Some of these may only occur to a very small extent and other reactions may also be occurring that are not included. Standard electrode potentials are shown:

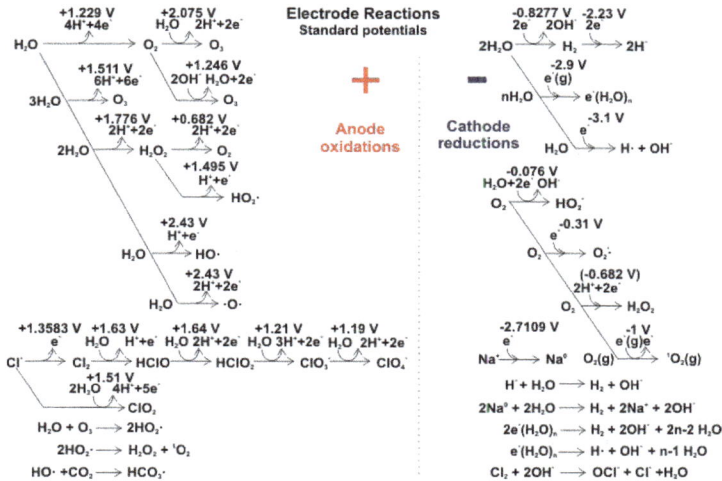

Electrolysis Compartments

Right is given a representation of the compartments in the electrolytic cell with some of their constituent molecules, ions, and radicals. Other materials may be present and some of the materials given may be at very low concentrations and have short half-lives.

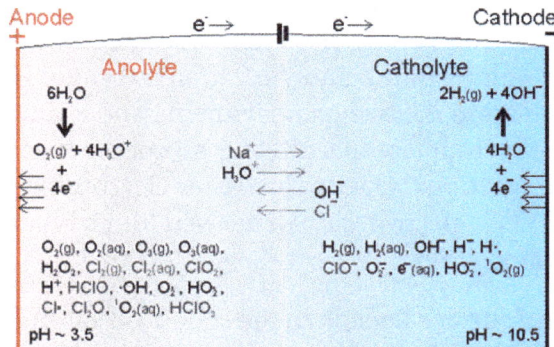

Electrode Compartment Contents in Water (NaCl) Electrolysis

Ozone, O_3

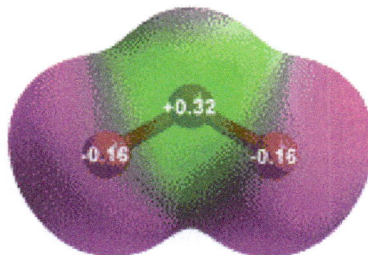

Important amongst the side products is ozone (O_3, see left).

The relative amount of O_3 produced (relative to molecular oxygen) depends on the overpotential, pH, radicals present and anode material. O_2 evolution is greater than that for O_3 due to the lower potential required. At low overpotentials, very little O_3 may be produced but at high current densities and overpotential, up to a sixth of the oxidized molecules may be O_3. As O_3 is more much more soluble than O_2, there may twice the dissolved O_3 than O_2 but the bubble gas will contain about 20 times the O_2 than O_3. Tin oxide anodes have proved useful for the production of O_3, particularly if doped with Sb and Ni, as they bind both oxygen molecules and hydroxyl radicals to facilitate the O_3 production. Ozone decomposes in water over a period of a few minutes. Decomposition of ozone (particularly in alkaline solution) gives rise to several strong oxidants including hydroxyl radicals (OH); an extremely strong oxidizer capable of killing viruses, amoebae, algae and dangerous bacteria, such as MRSA and Legionella.

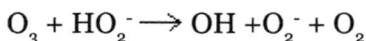

$$2O_3 \longrightarrow 3O_2$$

$$O_3 + OH^- \longrightarrow HO_2^- + O_2$$

$$O_3 + HO_2^- \longrightarrow OH + O_2^- + O_2$$

Although charged ions are attracted into the compartments by virtue of the applied potential, oppositely charged ions are created in both compartments due to the electrolytic reactions. Thus, for example, Na^+ ions enter the catholyte from the anode compartment but excess OH^- is produced at the same time at the cathode. The concentration of the OH^- ions will be generally expected to be greater than the increase in cations in the catholyte and the concentration of the H_3O^+ ions will be generally expected to be greater than any increase in anions in the anolyte. Often a conductive but semi-permeable membrane is used to separate the two compartments and reduce the movement of the products between the electrode compartments; a process that improves the yield by reducing back and side reactions. Due to the easier electrolysis of water containing ^1H rather than ^2H (D) or ^3H (T), electrolysis can be used for producing water with reduced or enriched isotopic composition.

Local inhomogeneities of surface tension in the produced gas bubbles may be caused by temperature or altered material concentration gradients at the interface. The resulting solute currents enhance the mass transfer and bubble growth.

When electrolysis uses short voltage pulses of alternating polarity at above 100 kHz the nanobubbles produced contain both H_2 and O_2 gases that can spontaneously react (combust) to form water whilst producing pressure jumps.

Although much time has been spent on investigating and modeling the electrolytic system, it is still not entirely clear how water is arranged at the surface of the electrodes. Alignment of the water dipoles with the field is expected, together with the consequential breakage of a proportion of the water molecules' hydrogen bonds. When the electrode processes occur it appears that singly-linked hydrogen atoms and singly-linked oxygen atoms are bound to the platinum atoms at the cathode and anode respectively. The binding energies of these hydrophilic intermediates are strongly influenced by hydrogen bonding (HB) to surface water molecules and the electrode composition. These bound atoms are able to diffuse around in two dimensions on the surface of their respective electrodes until they take part in their further reaction. Other atoms and polyatomic groups may also bind similarly to the electrode surfaces and subsequently undergo reactions. Molecules such as O_2 and H_2 produced at the surfaces may enter nanoscopic cavities in the liquid water (nanobubbles) as gases or become solvated by the water.

Proposed Mechanism for Electrolysis on Platinum

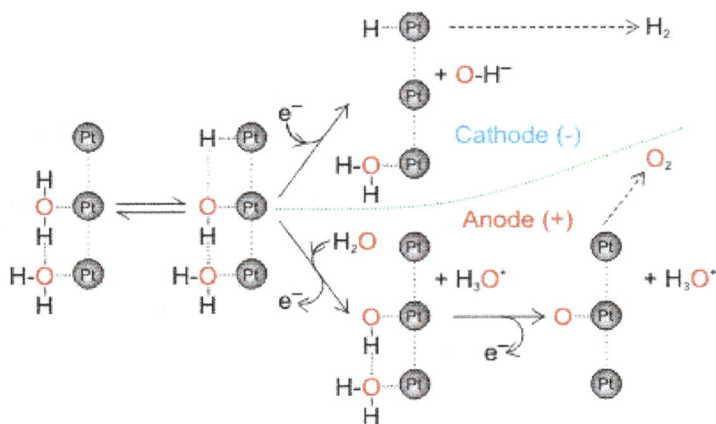

Gas-containing cavities in liquid solution grow or shrink by diffusion according to whether the solution is over-saturated or under-saturated with the dissolved gas. Given suitable electrodes, the size of the cathodic hydrogen bubbles depends on the overvoltage with nanobubbles being formed at low overvoltages and larger bubbles being formed at higher overvoltages. Larger micron-plus sized bubbles have sufficient buoyancy to rise through the solution and release contained gas at the surface before all the gas dissolves. With smaller bubbles a pressure is exerted by the surface tension is in inverse proportion to their diameter and they may be expected to collapse. However, as the nanobubble gas/liquid interface is charged, an opposing force to the surface tension is introduced, so slowing or preventing their dissipation. Electrolytic solutions have been proven to contain very large numbers of gaseous nanobubbles. The 'natural' state of such interfaces appears to be negative. Other ions with low surface charge density (such as Cl^-, ClO^-, HO_2^- and O_2^-) will also favor the gas/liquid interfaces as probably do hydrated electrons. Aqueous radicals also prefer to reside at such interfaces. From this known information it seems clear that the nanobubbles present in the catholyte will be negatively charged but those in the anolyte will probably possess little charge (with the produced excess positive H_3O^+ ions canceling out the natural negative charge). Therefore, catholyte nanobubbles are not likely to lose their charge on mixing with the anolyte stream and are otherwise known to be stable for many minutes. Additionally, gas molecules may become charged within the nanobubbles (such as the superoxide radical ion, O_2^-), due to the decay of ozone present and the excess potential on the cathode, increasing the overall charge of the nanobubbles and, probably, the stability of that charge. The raised temperature at the electrode surface, due to the excess power loss over that required for the electrolysis, may also increase nanobubble formation by reducing local gas solubility. Clearly increasing the pressure on solutions containing nanobubbles will also slow down their dissipation if this pressure has increased the dissolved gas content.

Sunlight, as an external electric field in water electrolysis, has proven to increase hydrogen production. This has been associated with an effect on the surface tension.

Commercial Systems

Commercial systems are more complex relative to the above descriptions. They must be safe, efficient and cheap to run. The electrodes must reduce the overpotentials required whilst keeping their capital costs low. The electrolytes must be clear of impurities that may poison the electrode surfaces and usually consist of strong alkali or acid. As the efficiency of an electrolyzer improves as the temperature increases, industrial electrolyzers run warm to hot. The best electrolyzers operate at 70-80% electricity-to-hydrogen efficiency and produce high-purity ($\approx 99.9\%$) hydrogen at ≈ 1 MPa pressure while providing intrinsically safe operation at all times.

Grotthuss Mechanism

It is generally thought that protons and hydroxide ions rapidly diffuse in liquid water, with protons diffusing almost twice as fast as hydroxide ions (and seven times as fast as Na+ ions). However, it should be recognized that these diffusivities are determined from movement in an electric field (at 100 v m^{-1}; H^+ and OHhave mobilities of 36.23 and 20.64 $\mu m\ s^{-1}$ respectively at 298 K)a , where the special mechanisms described below are operational, and the true diffusive movements of the ions may be somewhat less (particularly as they are attached to their attendant hydrogen-bonded water and accompanied by thier counter ions), as can be recognized by the proton diffusional limitations that take place at the surface of some immobilized enzymes.

Diffusion of Hydrogen Ions

The Grotthuss mechanism, whereby protons tunnel from one water molecule to the nextb via hydrogen bonding, is the usual mechanism given for facilitated proton mobility. The process is similar to that of autodissociation; the mechanism causing the ions (H^+, OH^-) to initially separate. Both processes increase with increasing temperature.

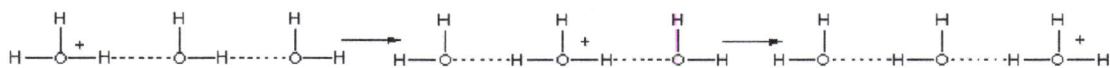

It is noteworthy that this process, although faster than translational diffusion, proves to be much slower than might be expected from its mechanism. This relative sluggishness may be due to the rotation of molecules required for trains of sequential proton movement and the consequential necessity for the breakage of hydrogen bonds. The strange effect of degassing increasing proton diffusion over ten-fold, however, indicates that the non-polar dissolved gas molecules, naturally present, disrupt the linear chains of water molecules necessary for the Grotthuss mechanism and so slow the proton movement. Over short hydrogen- bonded water wires there can be correlated movement of protons due to quantum fluctuations and extremely rapid transit through the existing suitably hydrogen bonded water wires. However, after a proton has moved along a chain of water molecules (in effect if not in body, by the excess proton disappearing at one end and appearing at the other end), it is clear that further proton movement requires a reorientation of the hydrogen bonding, if continued proton tunneling through the same molecules and in the same direction is to proceed.

In order to migrate, the ions must be associated with hydrogen bonded clusters; the stronger and more extensive the cluster, the faster the migration. Stronger hydrogen bonding causes the O···O distance to be shorter, so easing the close approach required for transfer. A limiting factor in the mobility for both ions is the breakage of an outer shell hydrogen bonds. This enables the proton to transfer from H_3O^+ and involves the additional energy requirements of stretching the outer hydrogen bonds due to the contraction of the O···O distance.

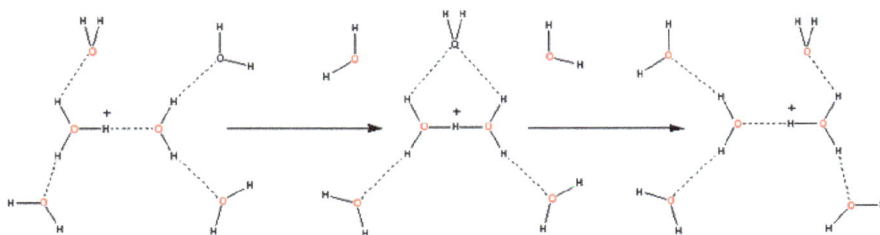

Oxonium Ion Transport Mechanism

The triangular arrangement of water molecules formed during proton transfer, has also been found in the protonated trimer ($H_7O_3^+$), and necessarily involves a rotation around the hydrogen bond as the 'Zundel' dihydronium ($H_5O_2^+$) ion flattens from its normal tetrahedral structure. The presence of the fourth water molecule associated with the $H_9O_4^+$ cluster is seen in a neutron diffraction study as oriented but distant (3.2 Å). Proton transport may also occur using 'Zundel' dihydronium ($H_5O_2^+$) ions only, as below, which involves the concerted movement of two molecules. Such proton jumps may be short or long. An ab initio simulation favored this mechanism, where $H_5O_2^+$ mobility was induced by thermal movement in the second solvation shell. An external electrical field was found to ease the process by suitably orienting the water in this direction. It has been suggested that proton mobility above 149° C decreases due to the decreasing amounts of $H_5O_2^+$ present.

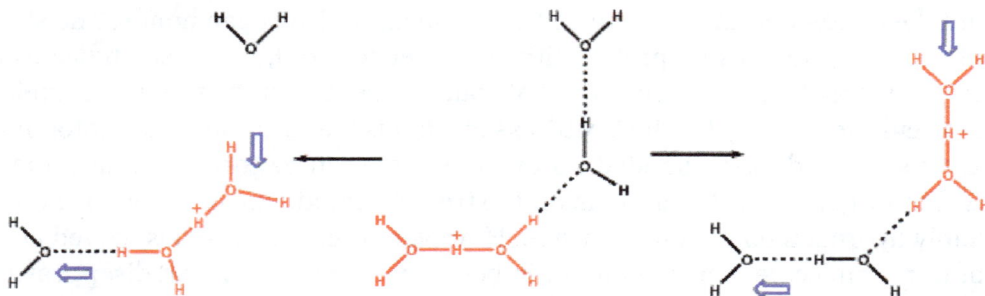

An additional and alternative mechanism has been proposed, using ab initio simulations but in agreement with the Zundel' dihydronium ($H_5O_2^+$) ions concentrations, by which the rapid diffusion of hydrogen ions, at temperatures below about 400° C, is due to the high diffusion of these $H_5O_2^+$ ions, allowed by the weaker surrounding hydrogen-bonded water network. Proton transport in water, protein channels and bioenergetic proteins has recently been reviewed. It is interesting that aquaporin water channels deliberately re-orient water molecules to preclude sequential hydrogen bonding so preventing proton transfer by the Grotthuss mechanism.

Diffusion of Hydroxyl Ions

A similar process to that for hydrogen ions was initially proposed for hydroxide mobility:

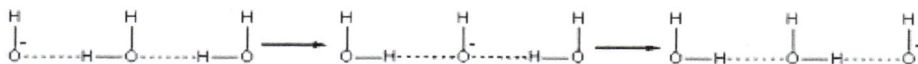

However it is now thought that hydroxide ions make use of an entirely different mechanism for diffusion in an electric field. It has been proposed that the movement of the hydroxyl ion is accompanied by a hyper-coordinating (that is, a fourth hydrogen bond donor) water molecule. The hydrated hydroxide ion is coordinated to four electron-accepting water molecules such that when an incoming electron-donating hydrogen bond forms (necessitating the breakage of one of the original hydrogen bonds) a fully tetrahedrally coordinated water molecule may be easily formed by the hydrogen ion transfer. The structure below left, HO- $(\cdot\cdot HOH)_4$, together with the more distant oriented water molecule below it, has been seen using neutron diffraction, with empirical structure refinement, of concentrated NaOH solutions. The different mechanism, involving extra hydrogen bond rearrangements plus re-orientations, is the reason for the reduced mobility of the hydroxide ion compared with the oxonium ion. Interestingly, the transfer involves an anionic trimer ($H_5O_3^-$), whereas hydrogen ion movement involved the cationic trimer ($H_7O_3^+$).

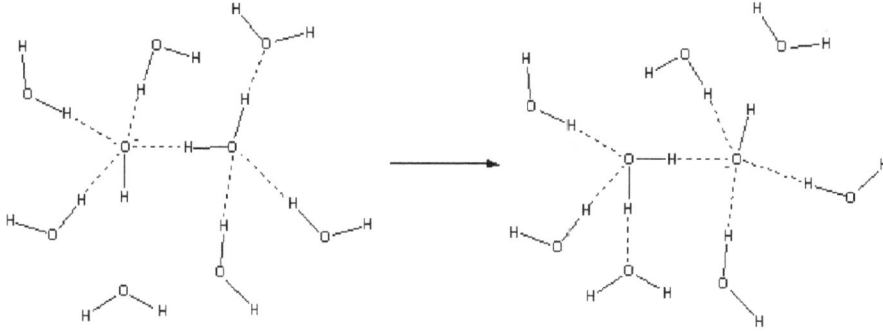

pH, Alkalinity and Buffering Capacity of Water

pH is an index of the concentration of hydrogen ions (H^+) in the water. It is defined as $-Log$ (H^+). The higher the concentration of hydrogen ions in the water, the lower the pH value is.

The pH scale ranges from 0 to 14 where:

- Water with a pH lower than 7 is considered to be acidic (higher H^+ concentration)

- Water with a pH higher than 7 is considered to be basic (lower H^+ concentration)

- Water with a pH of 7.0 is considered to be neutral

Since pH is on a logarithmic scale, a change of one unit in the pH (e.g. from 5.0 to 6.0) means a 10 fold change in the concentration of H^+ ions.

The hydrogen ions take part in most of the chemical reactions present in water and soil, which makes them extremely important. Their concentration (hence, the pH) influences the solubility of fertilizers, the ionic forms of salts (e.g. PO_4^{-3} vs $H_2PO_4^-$), the availability of nutrients to plants, stability of chelates etc.

A water or a soil solution with a pH that is too high can result in nutrient deficiencies, mainly micronutrients such as iron. Keeping the pH of the irrigation water below 7.0 is also important in order to prevent emitter clogging due to sedimentation of salts.

On the other hand, pH that is too low might result in micronutrient toxicities and damage to the plant's root system.

The desirable pH range in the root zone that is comfortable for most plants is 5.5-6.5. Therefore, many growers have to add acid to their irrigation water. Adding acid actually means adding hydrogen ions. However, to determine the amount of acid to be added, it is not enough to know the pH of the water. Another vital parameter must be taken into consideration: the water alkalinity.

Water Alkalinity and Buffering Capacity

The alkalinity of water is related to the pH, but it is actually a different parameter. It is a measure of the capacity of the water to resist changes in pH or, in other words, it is the buffering capacity of the water. Don't confuse "Alkalinity" with "Alkaline" (which means a pH of 7.0-14.0).

The main components of the water alkalinity are:

- Carbonates (CO_3^{-2})
- Bicarbonates (HCO_3^-)
- Soluble hydroxides (OH^-)

Alkalinity is usually expressed as ppm or mg/L of Calcium Carbonate ($CaCO_3$).

The higher the alkalinity, the more acid can be added without considerably changing the pH. This is because the bicarbonates (HCO_3^-) and carbonates (CO_3^{-2}) react with the hydrogen ions (H^+) contributed by the acid, preventing them from dropping the pH.

Once all the alkalinity components in the water are neutralized by the acid, the concentration of the free hydrogen ions in the water increases and there is a dramatic drop in the pH of the water. The following graph illustrates this "breaking point", where the pH drops:

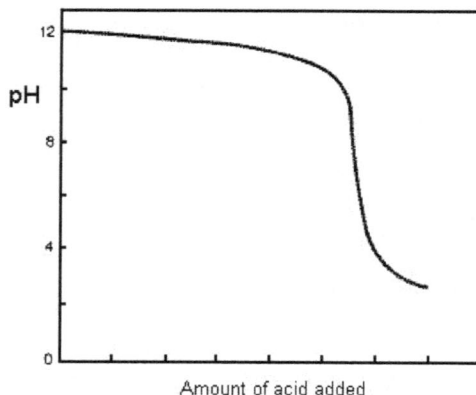

Here is simple example of how buffering capacity of the water influences daily decisions:

- Grower A has source water with a pH of 9.0 and alkalinity of 45 mg/L $CaCO_3$.

- Grower B has source water with a pH of 8.0 and alkalinity of 120 mg/L $CaCO_3$.

Both growers need the pH of their irrigation water to be 5.0, and use sulfuric acid for this purpose.

Even though Grower A's water has a higher pH (the concentration of hydrogen ions in his water is 10 times higher than Grower B's water) he actually needs to add less acid than grower B to reach the same target pH.

Less than 45 mg/L $CaCO_3$ is considered to be a low water alkalinity, with low buffering capacity. Acid added to this water will quickly affect its pH.

Therefore, it is obvious that both pH and alkalinity are essential for finding the correct amount of acid you have to add to the irrigation water in order to reach the required pH.

Here are the equations that relate pH to alkalinity:

$$pH = 6.37 + \log (HCO_3^-/H_2CO_3) \quad \text{or} \quad pH = 10.33 + \log (CO_3^{-2}/HCO_3^-)$$

References

- R. G. Wilkins Kinetics and Mechanism of Reactions of Transition Metal Complexes, 2nd Edition, VCH, Weinheim, 1991. ISBN 1-56081-125-0

- What-is-hydrolysis-375589: thebalance.com, Retrieved 31 June 2018

- H. Plieninger; Gunda Keilich (1956). "Die Dienol-Benzol-Umlagerung" [The dienol-benzene re-arrangement]. Angew. Chem. (in German). 68 (19): 618. doi:10.1002/ange.19560681914

- Hydrolysis-of-salts-equations: lumenlearning.com, Retrieved 19 May 2018

- "Investigation of pollution incidents". Queensland Government - Department of Environment and Heritage Proetection. 21 July 2016. Retrieved 1 August 2016

- Hydrolysis-in-acid-base-reactions: study.com, Retrieved 11 March 2018

- Cotton, F. A.; Fair, C. K.; Lewis, G. E.; Mott, G. N.; Ross, F. K.; Schultz, A. J.; Williams, J. M. (1984). "Precise Structural Characterizations of the Hexaaquovanadium(III) and Diaquohydrogen Ions. X-ray and Neutron Diffraction Studies of [V(H2O)6][H5O2](CF3SO3)4". Journal of the American Chemical Society. 106: 5319–5323. doi:10.1021/ja00330a047

- Water-of-crystallization: freechemistryonline.com, Retrieved 21 June 2018

5

Water Cycle

The mass of water remains more or less constant over time on Earth. Water moves from one reservoir to the other due to physical processes of evaporation, precipitation, condensation, surface runoff, etc. This is an important chapter, which will analyze in detail about the water cycle and the physical processes influencing the water cycle.

Hydrological Cycle

Water is most commonly found in its liquid form, in rivers, oceans, streams, and in the earth. The sun's rays constantly warm the water found in these places and, whether through this heat or through man-made means, the water particles gain energy and spread, turning the water from a liquid into a vapor through evaporation. The water vapor, thus becoming less dense, rises with the warm air into the sky where it sticks to other water particles to form clouds.

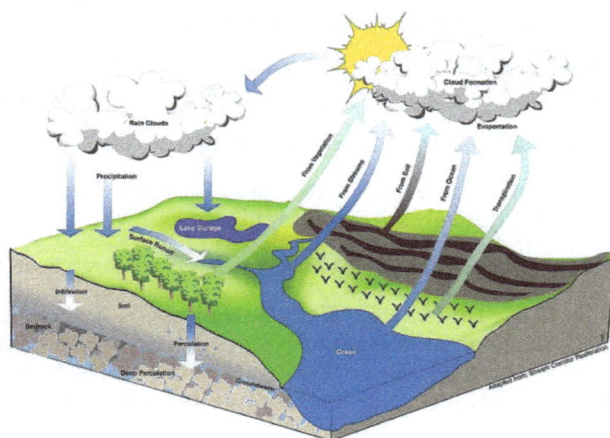

The Hydrologic Cycle

Typically, we consider the boiling point of water to be a hundred degrees centigrade, which is certainly true when pressure and humidity are normal. However, places such as mountains, where humidity is low and pressure is even lower, require less energy to boil away the water.

Along with the water vapor, some small particles can often rise up to form clouds. It is not only liquid water that can evaporate to become water vapor, but ice and snow, too. This process is simple enough, however there are a few things to note about evaporation.

This simple explanation, however, does not do justice to the complexity of the hydrologic cycle, which comprises many more steps. Here is a breakdown of the different steps of the hydrologic cycle.

Different Steps of the Hydrologic Cycle

Here is a breakdown of the different steps of the hydrologic cycle.

Water is most commonly found in its liquid form, in rivers, oceans, streams, and in the earth. The sun's rays constantly warm the water found in these places and, whether through this heat or through man-made means, the water particles gain energy and spread, turning the water from a liquid into a vapor through evaporation. The water vapor, thus becoming less dense, rises with the warm air into the sky where it sticks to other water particles to form clouds.

- Evaporation: is frequently used as a catch-all term to refer to the process of water turning to water vapor, however there is another distinct term for the evaporation of water from a plant's leaves.

- Evapotranspiration: makes up a large portion of the water in the planet's atmosphere due to the sheer surface area of the globe covered by flora. The majority of water in the atmosphere comes from lakes and oceans – around ninety per cent – but in terms of land-based water, evapotranspiration is an important player.

- Sublimation: as the process is called, results from when pressure and humidity are low as noted above. It is not only liquid water that can evaporate to become water vapor, but ice and snow, too. Due to lower air pressure, less energy is required to sublimate the ice into vapor. Other factors which can aid in sublimation are high winds and strong sunlight, which is why mountain ice is a prime candidate for sublimation, while ground ice sublimation is not so common. A good, visible example of sublimation is dry ice, which emits a thick layer of water vapor due to its lower energy requirement.

The further above sea level one gets, the cooler the air. When water vapor reaches this plane, it cools significantly and clumps together. So stuck together, this newly formed cloud is subject to the movement of the wind and the changes in the air pressure, which is what moves the water around the planet. There are a couple of things that can happen to the vapor in this state.

- Precipitation/Rainfall: refers to vapor that cools to any temperature above freezing point (zero degrees centigrade) will condense, becoming droplets of liquid water. These droplets form when the water vapor condenses around particles and other matter that rises up with the water during evaporation, giving a nucleus to the water droplet so that it can clump together. Once a number of these tiny, particle based droplets form, they collide and clump together as larger droplets. At a certain point, the droplet will become big enough that its mass will be subject to the force of gravity at a rate faster than the force of the updraft in the air around it. At this point, the water falls to earth.

- Snow: refers to frozen water falling from the sky. When it is particularly cold or the air pressure is exceptionally low, these water droplets will crystalize before falling.

- Sleet: is a bitterly cold, half-frozen slush. This third state occurs when the conditions are not quite cold enough to keep the crystals frozen and the water either does not freeze fully or if precipitation occurs in particularly cold conditions, or conditions in which the air pressure is very low, then these water droplets can quite often crystallize and freeze. This causes the water to fall as solid ice, known melts somewhat in the process.

When water falls to earth, it quite often ends up on tarmac or over man-made surfaces where it quickly evaporates again.

Infiltration: is water that doesn't evaporate after precipitation and falls into soil and other absorbent surfaces. The water moves throughout the soil, saturating it.

Groundwater Storage: is water that has not precipitated or run off into streams or rivers, but instead moves deep underground forming pools known as "groundwater storage". In groundwater storage, water joins up in the soil and forms pools of saturated soil instead of escaping the soil. These pools are called "aquifers".

- Springs: occur when an aquifer becomes oversaturated, and the excess water leaks out of the soil onto the surface. Most commonly, springs will emerge from cracks in rocks and holes in the ground. Sometimes, if conditions are particularly volcanic, the spring will heat up and form "hot springs".

- Runoff: After heavy rainfall has saturated the soil it will cease to absorb water and additional rainfall, as well as melted snow and ice, will simply flow off of the surface. The flow follows gravity down hills, mountains, and other inclines to form streams and join rivers. This is known as "runoff", and it is the principle way in which water moves along the Earth's surface. The rivers and streams are pulled by gravity until they pool together to form lakes and oceans.

- Streamflow: is the direction the runoff takes to form a stream and it is this flow which dictates the river's currents depending on how close they are to the ocean. Because ice and snow make up a large portion of the water involved in runoff, heatwaves are a principle cause of flooding as the water stored on the surface is suddenly released into runoff flow. In particular, a warm spring following a cold winter can result in quite spectacular flood, as a large volume of water gets stored in ice and snow only to quickly melt and form new streams.

- Ice Caps: occur when a large volume of snow falls and is not evaporated or sublimated, the ice compacts under its own weight to form these caps. Ice caps, glaciers, and ice sheets contain a huge amount of water, and those found in the polar regions of the planet are the largest stores of ice found in the world. As the atmosphere warms up slowly, more and more of this ice melts and evaporates, releasing more water into the hydrologic cycle. It is this process which causes rises in the ocean levels.

The hydrologic cycle happens continuously, with all different steps happening simultaneously around the world. The biggest concern that many have with the hydrologic cycle is the availability of drinkable water, which is something that is constantly in flux, and the melting of the huge ice storage sheets at the polar caps. Having an understanding of the different steps of the hydrologic cycle is an important step in understanding what effect human activity has on the world's water.

Water Vapor

Water vapor, water vapour, also aqueous vapor, is the gas phase of water. There is no difference between the terms gas and vapor, but gas is used commonly to describe a substance that appears in the gaseous state under standard conditions of pressure and temperature, and vapor to describe the gaseous state of a substance that appears ordinarily as a liquid or solid. Because water by definition is a liquid, when used in a direct context, the gas phase of water is referred to as water vapor.

Water vapor is not visible, therefore clouds, fog and most other formations within the atmosphere that can be seen by the naked eye are not water vapor. Water vapor, however, can be sensed. If enough of it is in the air it is felt as humidity. Water vapor is one state of the water cycle within the hydrosphere. Water vapor can be produced from the evaporation of liquid water or from the sublimation of ice. Under normal atmospheric conditions, water vapor is continuously generated by evaporation and removed by condensation.

Water vapor is vital to weather and climate as clouds, rain and snow have their source in water vapor. All of the water vapor that evaporates from the surface of the Earth eventually returns as precipitation - rain or snow. Water vapor is also the Earth's most important greenhouse gas, giving us over 90% of the Earth's natural greenhouse effect, which helps keep the Earth warm enough to support life. When liquid water is evaporated to form water vapor, heat is absorbed. This helps to cool the surface of the Earth. This "latent heat of condensation" is released again when the water vapor condenses to form cloud water. This source of heat helps drive the updrafts in clouds and precipitation systems. In order to understand water vapor, some insight must be given into water.

Water and its Properties

Water is one of the main sources of the energy needed to run the Earth's weather machine. All substances, including water exist in three phases: Solid, liquid, and gas. No matter what phase it's in, a water molecule is the same - an atom of oxygen with two atoms of hydrogen attached. Their energy, as exhibited by the speed of their movement, is the only thing that makes millions of water molecules act differently when they get together in water's different phases. Water's phase changes help drive the weather as each change either releases or takes up energy in the form of latent heat. The energy of the molecules known as the average speed of molecular motion, determines the phase of a substance. Temperature, in turn, is a measure of the average speed of molecular motion. Water is special because it's the only substance that can exist in all three phases at Earth's ordinary temperatures, and it's common to have all three phases together at the same time. To understand the weather, you need to understand what happens when water changes its state. These changes are:

Evaporation: From liquid to gas (water vapor).

- Condensation: From gas (water vapor) to liquid.

- Freezing: From liquid to solid (ice).

- Melting: From solid to liquid.

- Deposition: From gas directly to solid without becoming liquid.

- Sublimation: From solid directly to gas.

The conversation of energy is one of the very basic laws of nature. It says that energy cannot be created or destroyed, but it can change form. This means that when water molecules slow down enough to change from vapor to liquid or ice, the kinetic energy of their movement changes into another form of energy, heat. When water evaporates from a pond to become water vapor, heat energy becomes the kinetic energy of the added motion.

Latent Heat When heat is added or extracted from a system the temperature changes. However, there are certain situations when the addition or subtraction of heat from a system does not result in a temperature change. In these situations, the heat that is added or extracted is being converted into energy to cause a phase change from one physical state to another.

Latent heat is the energy that is required to change a substance from one form to another. It is the enthalpy change that accompanies a phase change at constant temperature and pressure. It is called latent because it is somewhat hidden from detection except for the physical change in a substance's form. The value of latent heat is dependent upon the exact nature of the phase change as well as on the specific properties of the substance. Latent heat is usually expressed in terms of calories per gram. The latent heat of water is 79.7cal/g

There are various forms of latent heat, depending on what transformation is occurring when it is taken up or released. These kinds are listed below with the amounts of energy involved in each. The figures below are those normally found in meteorology texts and are for temperatures found in the atmosphere, such as 0 Celsius (32 F). Latent heat of condensation (Lc): Refers to the heat gained by the air when water vapor changes into a liquid. Lc=2500 Joules per gram (J/g) of water or 600 calories per gram (cal/g) of water. Latent heat of fusion (Lf): Refers to the heat lost or gained by the air when liquid water changes to ice or vice versa. Lf=333 Joules per gram (J/g) of water or 80 calories per gram (cal/g) of water. Latent heat of sublimation (Ls): Refers to the heat lost or gained by the air when ice changes to vapor or vice versa.

Ls=2833 Joules per gram (J/g) of water or 680 calories per gram (cal/g) of water. Latent heat of vaporization (Lv): Refers to the heat lost by the air when liquid water changes into vapor. This is also commonly known as the latent heat of evaporation. Lv= -2500 Joules per gram (J/g) of water or -600 calories per gram (cal/g) of water.

Latent Heat of Vaporization

The latent heat of a physical transformation from one phase to another can also be thought of as the amount of energy required to rearrange the molecules of a substance. When a solid is transformed into a liquid, the magnitude of the vibrations of the atoms about their equilibrium positions becomes large, large enough to overcome the attractive forces that bind the atoms together into a solid form. The latent heat is the energy required to break these bonds and transform the material from the ordered solid state to the disordered liquid state. Just as energy is required to break the bonds binding the atoms of a solid together, energy is also required to weaken the attractive forces between the molecules in a liquid in order for it to become a vapor. In a liquid the molecules are closer together than in a vapor phase and so the forces between them are stronger than in a gas. In order to separate the molecules the attractive forces must be broken. Since the average distance between molecules in a vapor are larger than either the liquid or the solid states, it is obvious that more work is required to separate the molecules to form a vapor. This explains why the latent heat of vaporization is much higher than the latent heat of fusion for a given substance.

The latent heat of evaporation or, vaporization is the energy process directly involved in the formation of water vapor. When heat is added to a liquid at its boiling point , with the pressure kept constant, the molecules of the liquid acquire enough energy to overcome the intermolecular forces that bind them together in the liquid state, and they escape as individual molecules of vapor until the vaporization is complete. Vaporization at the boiling point is known simply as boiling. The temperature of a boiling liquid remains constant until all of the liquid has been converted to a gas.

For each substance a certain specific amount of heat must be supplied to vaporize a given quantity of the substance. The quantity of heat applied for each gram (or each molecule) undergoing the change in state depends on the substance itself. For example, the amount of heat necessary to change one gram of water to steam at its boiling point at one atmosphere of pressure, i.e., the heat of vaporization of water, is approximately 540 calories

Evaporation

Liquids can also change to gases at temperatures below their boiling points. Vaporization of a liquid below its boiling point is called evaporation, which occurs at any temperature when the surface of a liquid is exposed in an unconfined space. When, however, the surface is exposed in a confined space and the liquid is in excess of that needed to saturate the space with vapor, an equilibrium is quickly reached between the number of molecules of the substance going off from the surface and those returning to it. A change in temperature upsets this equilibrium; a rise in temperature, for example, increases the activity of the molecules at the surface and consequently increases the rate at which they fly off. When the temperature is maintained at the new point for a short time, a new equilibrium is soon established.

Vapor Pressure

All liquids and solids have a tendency to evaporate to a gaseous form, and all gases have a tendency to condense back into their original form (either liquid or solid). At any given temperature, for a particular substance, there is a pressure at which the gas of that substance is in dynamic equilibrium with its liquid or solid forms. This is the vapor pressure of that substance at that temperature. The vapor pressure of a liquid is the pressure exerted by its vapor when the liquid and vapor are in dynamic equilibrium.

Vapor pressures differ for different substances at any given temperature, but each substance has a specific vapor pressure for each given temperature. At its boiling point the vapor pressure of a liquid is equal to atmospheric pressure. For example, the vapor pressure of water, measured in terms of the height of mercury in a barometer, is 4.58 mm at $0°$ C and 760 mm at $100°$ C (its boiling point).

Vapor pressures increase with temperature. The vapor pressure of any substance increases non-linearly with temperature according to the ClausiusClapeyron relation : $\ln(P_2/P_1) = -DH_{vap}/R * (1/T_2 - 1/T_1)$. The most common unit for vapor pressure is the torr. 1 torr = 1 mm Hg (one millimeter of mercury).

Most materials have very low vapor pressures. Water has a vapor pressure of approximately 15 torr at room temperature. But because vapor pressures increase with temperature; water will have a vapor pressure of 760 torr = 1 atm at its boiling point of 100°C (212°F). Conversely, vapor pressure decreases as the temperature decreases.

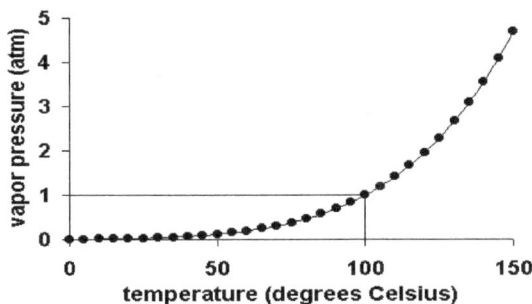

The equilibrium vapor pressure is an indication of a liquid's evaporation rate. It relates to the tendency of molecules and atoms to escape from a liquid or a solid. A substance with a high vapor pressure at normal temperatures is often referred to as volatile.

Water Vapor in the Earth's Atmosphere

The troposphere contains 75 percent of the atmosphere's mass—on an average day the weight of the molecules in air is 1.03 kg/sq cm (14.7 lb/sq in)—and most of the atmosphere's water vapor. Water vapor varies by volume in the atmosphere from a trace, or 0% to about 4%. Therefore, on average, only about 2 to 3% of the molecules in the air are water vapor molecules. The amount of water vapor in the air is small in extremely arid areas and in location where the temperatures are very low (i.e. polar regions, very cold weather). The volume of water vapor is about 4% in very warm and humid tropical air.

The amount of water vapor in the air cannot exceed 4% because temperature sets a limit to how much water vapor can be in the air. Even in tropical air, once the volume of water vapor in the atmosphere approaches 4% it will begin to condense out of the air. The condensing of water vapor prevents the percentage of water vapor in the air from increasing. If temperatures were much warmer, there would be a potential to have more than 4% water vapor in the atmosphere.

The concentration of water vapor in the atmosphere reflects the number of molecules of water compared with the total number of air molecules (mainly nitrogen and oxygen). Humidity is a measure of the amount of water vapor in the air. One way to represent humidity is the mixing ratio, defined as the mass of water vapor "mixed with" each unit mass of air. The mixing ratio is usually expressed as the number of grams of water vapor in each kilogram of air. In the atmosphere, the mixing ratio can vary from nearly zero (in deserts and polar regions and at high altitudes) to as much as 30 grams per kilogram (in warm, moist tropical regions). Other measurements of humidity include the relative humidity, which reflects the ratio of the actual pressure of water vapor in a sample of air to the pressure necessary to saturate that air at a given temperature and dew point temperature, the temperature to which the air must be cooled for water vapor to reach saturation.

Water Vapor and the Greenhouse Effect

In a very rough approximation the following trace gases contribute to the greenhouse effect: 60% water vapor, 20% carbon dioxide (CO_2). The rest (~20%) is caused by ozone (O_3), nitrous oxide (N_2O), methane (CH4), and several other species. Water vapor amplifies the anthropogenic contribution to greenhouse warming through a positive feedback. This amplification is counteracted by the increased reflection off clouds. Water vapor is known to be Earth's most abundant greenhouse gas, but the extent of its contribution to global warming has been debated. Using recent NASA satellite data, researchers have estimated more precisely than ever the heat-trapping effect of water in the air, validating the role of the gas as a critical component of climate change. Andrew Dessler and other scientists at Texas A and M University in College Station confirmed that the heat-amplifying effect of water vapor is potent enough to double the climate warming caused by increased levels of carbon dioxide in the atmosphere.

With new observations, the scientists confirmed experimentally what existing climate models had anticipated theoretically. The research team used novel data from the Atmospheric Infrared Sounder (AIRS) on NASA's Aqua satellite to measure precisely the humidity throughout the lowest 10 miles of the atmosphere.

That information was combined with global observations of shifts in temperature, allowing researchers to build a comprehensive picture of the interplay between water vapor, carbon dioxide, and other atmosphere-warming gases. Most agree that if you add carbon dioxide to the atmosphere, warming will result. The amount of warming that occurs can be found by estimating the magnitude of water vapor feedback. Increasing water vapor leads to warmer temperatures, which causes more water vapor to be absorbed into the air. Warming and water absorption increase in a spiraling cycle.

Water vapor feedback can also amplify the warming effect of other greenhouse gases, such that the warming brought about by increased carbon dioxide allows more water vapor to enter the atmosphere. More water vapor in the air also gives rise to an increase in the formation of clouds in the troposphere. Clouds do consist of small water droplets, though, and, hence, they do absorb radiation. But they also have a moderating effect on the process of earth's warming because clouds reflect a significant portion of solar isolation. Thus, this portion does not reach the surface of the earth and thus surface is less heated.

Climate models have estimated the strength of water vapor feedback, but until now the record of water vapor data was not sophisticated enough to provide a comprehensive view of at how water vapor responds to changes in Earth's surface temperature. Past ground instruments and previous space-based could not measure water vapor at all altitudes in Earth's troposphere - the layer of the atmosphere that extends from Earth's surface to about 10 miles in altitude.

AIRS is the first instrument to distinguish differences in the amount of water vapor at all altitudes within the troposphere. Using data from AIRS, the team observed how atmospheric water vapor reacted to shifts in surface temperatures between 2003 and 2008. By determining how humidity changed with surface temperature, the team could compute the average global strength of the water vapor feedback.

"This new data set shows that as surface temperature increases, so does atmospheric humidity," Dessler said. "Dumping greenhouse gases into the atmosphere makes the atmosphere more humid. And since water vapor is itself a greenhouse gas, the increase in humidity amplifies the warming from carbon dioxide."

Specifically, the team found that if Earth warms 1.8 degrees Fahrenheit, the associated increase in water vapor will trap an extra 2 Watts of energy per square meter (about 11 square feet)."That number may not sound like much, but add up all of that energy over the entire Earth surface and you find that water vapor is trapping a lot of energy," Dessler said. "We now think the water vapor feedback is extraordinarily strong, capable of doubling the warming due to carbon dioxide alone."

Because the new precise observations agree with existing assessments of water vapor's impact, researchers are more confident than ever in model predictions that Earth's leading greenhouse gas will contribute to a temperature rise of a few degrees by the end of the century. "This study confirms that what was predicted by the models is really happening in the atmosphere," said Eric Fetzer, an atmospheric scientist who works with AIRS data at NASA's Jet Propulsion Laboratory in Pasadena, Calif. "Water vapor is the big player in the atmosphere as far as climate is concerned. (Hansen.K. Water Vapor Confirmed As Major Player In Climate Change).

Detecting Water Vapor in the Atmosphere

Water plays a crucial role in weather and climate, and identifying the amount of water vapor in the atmosphere will help scientists understand clouds, severe weather, precipitation, hydrology, and global climate change. There has been a general lack of information on the way water moves around in Earth's atmosphere - where it comes from and where it ends up. The details of this journey are critical for understanding clouds and climate, as well as changes in precipitation patterns and water resources. Because of water vapor's great mobility and brief residence time, water vapor is a central component of the global hydrological cycle. How this cycle may change globally and regionally in the future is a major issue for climate science and society. Water vapor is vital for Earth's energy and water cycles, it must be monitored in time and space if we are to explain and predict behavior of the climate system.

Measuring water vapor sufficiently well to properly understand the processes responsible for its variability has proven disappointingly elusive. This situation results in part because water vapor is not dynamically constrained, and its high special variability makes adequate sampling difficult. Problems associated with the various water vapor measurement technologies also have hindered progress. Standard humidity sensors carried out by radiosondes have complex error statistics up through the mid troposphere and their performance is severely diminished at higher levels. The network of radiosonde stations works best on ground locations and the number of these has diminished over the past decades. Satellite images provide global coverage, but their vertical resolution in the lower troposphere, where water vapor is most abundant, is poor compared with that of ground-based radiosondes systems. Long term water vapor coverage can also be problematic due to several factors such as gradual changes in instrument sensitivity and local crossing time, abrupt changes resulting from satellite replacements, and short or intermittent system lifetimes.

Evaporation

Evaporation in nature is one of the main components of the global water cycle and it is the only means of water vapor transport from land and ocean to the atmosphere. As much heat is absorbed during conversion of water into vapor, evaporation is also an important factor of heat exchange on the Earth's surface. The particular importance of evaporation as a geographic process lies in the fact that characteristics of evaporation rate are included into the fundamental equations of water and heat balances for the underlying surface, thus establishing a direct relationship between water exchange and heat exchange processes. Evaporation in Nature depends on many factors, the most important of which are moisture content in the underlying surface and meteorological conditions above the surface. Meteorological conditions comprise solar radiation intensity, air humidity and wind velocity. From the position of the molecular-kinetic theory, evaporation is the result of two interacting processes, i.e. take-off of quickly moving molecules from the surface of water, snow, ice, soil, droplets and crystals in the atmosphere, and return of some water vapor molecules, above the evaporating surface, back to liquid or solid state. Rise of temperature of the evaporating surface results in a higher rate of water molecule take-off and their transport to the atmosphere as water vapor. Accordingly, the evaporation rate increases, too. On the other hand, a higher water vapor concentration above the evaporating surface increases the volume of the returning molecules. In this case, evaporation rate decreases. During the description of water evaporation in Nature, the different stages of the process are often separated. If water on the surface is in its liquid state and rises to the atmosphere as water vapor, this stage is termed 'evaporation'. If the quantity of molecules coming to the underlying surface exceeds the quantity transported to the atmosphere (e.g. observed during dew formation), this stage is termed 'condensation'. If evaporation takes place directly from the surface of frozen water, and, vice versa on cooling from a gaseous state to solid without apparent liquefaction, it is termed 'sublimation'. Phase conversion from one state to another is accompanied by absorption or discharge of heat. For example, evaporation of a unit of water mass, i.e. liquid water conversion into gas is accompanied by heat absorption:

$$L_E = (25 - 0,027\theta) \cdot 10^5$$

LE is specific heat of evaporation in Joule/kg – the quantity of heat required to evaporate 1 kg of water; θ is the temperature of the evaporating surface, in $^\circ$ C. A phase conversion of water directly from a solid to a gaseous state (sublimation) is accompanied by absorption of additional heat corresponding to heat loss for melting.

Specific heat of melting a unit of distilled water mass at 0 $^\circ$ C L_m equals 33.5·x 10⁴ Joule/kg. Water freezing and condensation are accompanied by equivalent heat discharge.

Background

The great variety of global landscapes (geotopes), each with characteristic features of evaporation, explain different methodological approaches to the study of evaporation. In general, we can distinguish evaporation from a water surface, evaporation direct from the soil surface (physical evaporation), evaporation from the surface of plants (transpiration), and total evaporation (or evapotranspiration), which comprises both physical evaporation and transpiration. There is also a term "evaporativity" (or potential evapotranspiration) which is evaporation from an optimally wet land area with constant compensation of water loss in soil. Also, depending on the research objective, it is possible to separate background evaporation (mean weighted over area) and evaporation from particular surfaces, i.e. from water, snow, agricultural field, forest, swamp, meadow, etc. The very first attempts to describe evaporation were made in the seventeenth century. Edmund Halley during his experiments proved that water circulation in Nature was closely connected with evaporation. After Halley, the first measurements of evaporation were made by D. Dobson who observed evaporation from a water surface in the vicinity of Liverpool from 1772 to 1775, from which he estimated monthly and annual evaporation. G.W. Rickman contributed much to development of evaporation theory and developed several new designs of evaporation pans for laboratory experiments. He studied the dependence of evaporation and condensation on the difference between air and water temperatures and some other factors. The first mathematical description of evaporation was made by John Dalton in 1802, and is known as Dalton's law. According to Dalton's low, the rate of evaporation to a free atmosphere is estimated by the following formula:

$$E = a(e_0 - e)$$

where: α is a coefficient which depends on different factors not taken into account and affecting evaporation rate; e0 - is the viscosity of saturated water vapor; e - is actual water vapor viscosity. Dalton's law is valid for open water surfaces. Later, great contributions to the study of evaporation from a water surface, based on analysis of turbulent water vapor diffusion to the atmosphere and on analysis of heat balance, were made by Jeffreys, Leichtmann, Svedrup, Montgomery, Schmidt, Shuleikin, and others.

It is more difficult to estimate specific features of evaporation from land. Basic physical principles of this process were developed late in the nineteenth and early in the twentieth centuries. The very first studies in this field were aimed at development of methods for measurements and computation of evapotranspiration (evaporation from soil plus transpiration from plants). During the second half of the twentieth century several designs of soil evaporimeters were prepared; they were separated into three types, i.e. evaporimeters with extreme wetting, evaporimeters with isolated monolith without additional wetting, and evaporimeters with screens. Evaporimeters of the first type included the evaporimeter of Dorandt, an earth-filled metal cylinder with a connecting tube put into a tank with water. The second type of evaporimeters was designed by M.A. Rykachev in 1898, and the third type of evaporimeter was developed by V.I. Popov in 1928. In some hydrological studies early in the twentieth century a water balance method was suggested to estimate evapotranspiration from large territories. This method was used to prepare maps of mean annual evaporation for the USA, European territory of the former Soviet Union, and other regions. Simple empirical graphs and formulas for estimation of evapotranspiration were widely used that time; they were based on the ratios between evaporation and different meteorological factors. These were the graphs of Meyer (1928), Kuzin (1934), formulas of Rykachev (1925), Blaney and Morin (1942), and others. Lack of any serious physical substantiation of this approach resulted in large errors in evaporation measurements when empirical schemes were applied. Theoretical methods for estimation of evaporation from land surface were first developed in the middle of the twentieth century, much later than the methods for computation of evaporation from a free water surface. Thornthwaite and Holzmann (1939) took the idea of Schmidt about evaporation measurement of a vertical turbulent flux of water vapor above an underlying surface and derived a formula to calculate evaporation rate based on measurements of specific air humidity and wind velocity at two elevations. This was a prototype of the present formulas based on the account of water vapor diffusion in the turbulent atmosphere.

Calculation of evaporation by the heat balance method was described by Skvortsov, Albrecht and others. Another research trend to study evaporation from land was connected with studies of the effect of different factors on evaporation from soil and on transpiration of plants. At the end of the nineteenth century numerous qualitative relations between evaporation rate from soil and physical and mechanical soil properties, plant cover and some meteorological factors were established. The discovery of three stages of evaporation made by the Russian pedologist P.S. Kossovic was a great achievement in studying specific features of evaporation from soil. He established that when soil saturated with moisture began to dry, the rate of evaporation is about constant at some time. Then, at a certain soil moisture content, the rate of evaporation trends to a rapid decrease to a relatively low value. After this, the evaporation rate decreases slowly until complete desiccation of the soil. The conclusion made by Kossovich were later proved by the English agrophysicists Kean and Fisher, as well as by numerous experiments. In the nineteenth century the concept of transpiration coefficient for a plant was introduced; it included the ratio between water weight lost in transpiration and increment of dry matter weight for a specified time interval. It was first assumed that the transpiration coefficients for similar plant species were more or less constant. This concept was refuted early in the twentieth century by Brigs and Schantz and by other scientists who established the considerable variability of transpiration coefficients during variable meteorological conditions. This confirmed the idea of K.A. Timiriazev who had noted early in the 1890s that water loss for transpiration in usual conditions greatly exceeded the actual water requirements for plants and it was more a physical evil than a necessary physiological demand of the plant. Livingstone made an attempt to establish relations between the rate of evaporation from the leaves of plants and rate of evaporation from an evaporimeter. As a result of his experiments, Livingstone discovered that the ratio between evaporation rate from plants and evaporation from the evaporimeter (under similar meteorological conditions) varied greatly depending on the state of the plant and on environmental factors. These changes were explained by Livingstone by the mechanism of opening and closing of stomata (small pores in the surface of leaves) under the influence of drying of leaves, if evaporation was intensive. Livingstone's studies were later continued by Walter. Some hydrologists tried to discover empirical dependences of water loss for transpiration of different vegetative cover upon a season, and tried to estimate mean total water loss for transpiration for the whole vegetative period. These quantitative assessments were quite approximate because they were obtained on the basis of using methods for transpiration measurements which were not satisfactory from the physical viewpoint. The results of present geophysical studies of evaporation under natural conditions (to be discussed below) made it possible to establish basic specific features of natural evaporation and to develop a number of effective methods for computation of evaporation from different underlying surfaces. Meanwhile, it should be noted that despite the progress in this field, the available methods for computation of evaporation are less accurate than methods for computation of the other main water balance components (precipitation and runoff).

Evapotranspiration

Evaporation is the process whereby liquid water is converted to water vapor (vaporization) and removed from the evaporating surface. Water evaporates from a variety of surfaces, such as lakes, rivers, pavements, soils and wet vegetation.

Energy is required to change the state of the molecules of water from liquid to vapor. Direct solar radiation and, to a lesser extent, the ambient temperature of the air provide this energy. The driving force to remove water vapor from the evaporating surface is the difference between the water vapor pressure at the evaporating surface and that of the surrounding atmosphere. As evaporation proceeds, the surrounding air becomes gradually saturated and the process will slow down and might stop if the wet air is not transferred to the atmosphere. The replacement of the saturated air with drier air depends greatly on wind speed. Hence, solar radiation, air temperature, air humidity and wind speed are climatological parameters to consider when assessing the evaporation process.

Where the evaporating surface is the soil surface, the degree of shading of the crop canopy and the amount of water available at the evaporating surface are other factors that affect the evaporation process. Frequent rains, irrigation and water transported upwards in a soil from a shallow water table wet the soil surface. Where the soil is able to supply water fast enough to satisfy the evaporation demand, the evaporation from the soil is determined only by the meteorological conditions. However, where the interval between rains and irrigation becomes large and the ability of the soil to conduct moisture to pear the surface is small, the water content in the topsoil drops and the soil surface dries out. Under these circumstances the limited availability of water exerts a controlling influence on soil evaporation. In the absence of any supply of water to the soil surface, evaporation decreases rapidly and may cease almost completely within a few days.

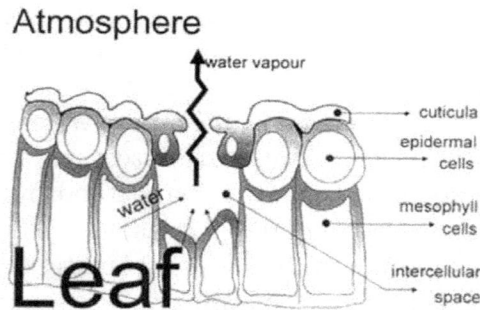

Figure: Schematic representation of a stoma.

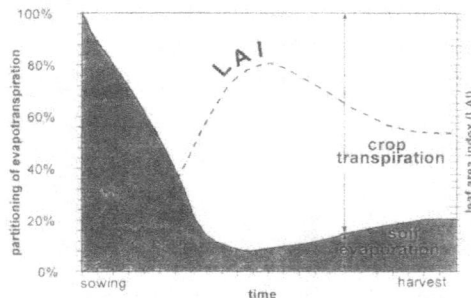

Figure: The partitioning of evapotranspiration into evaporation and transpiration over the growing period for an annual field crop.

Transpiration consists of the vaporization of liquid water contained in plant tissues and the vapor removal to the atmosphere. Crops predominately lose their water through stomata. These are small openings on the plant leaf through which gases and water vapor pass. The water, together with some nutrients, is taken up by the roots and transported through the plant. The vaporization occurs within the leaf, namely in the intercellular spaces, and the vapor exchange with the atmosphere is controlled by the stomatal aperture. Nearly all water taken up is lost by transpiration and only a tiny fraction is used within the plant.

Transpiration, like direct evaporation, depends on the energy supply, vapor pressure gradient and wind. Hence, radiation, air temperature, air humidity and wind terms should be considered when assessing transpiration. The soil water content and the ability of the soil to conduct water to the roots also determine the transpiration rate, as do waterlogging and soil water salinity. The transpiration rate is also influenced by crop characteristics, environmental aspects and cultivation practices. Different kinds of plants may have different transpiration rates. Not only the type of crop, but also the crop development, environment and management should be considered when assessing transpiration.

Evapotranspiration

Evaporation and transpiration occur simultaneously and there is no easy way of distinguishing between the two processes. Apart from the water availability in the topsoil, the evaporation from a cropped soil is mainly determined by the fraction of the solar radiation reaching the soil surface. This fraction decreases over the growing period as the crop develops and the crop canopy shades more and more of the ground area. When the crop is small, water is predominately lost by soil evaporation, but once the crop is well developed and completely covers the soil, transpiration becomes the main process. In Figure above the partitioning of evapotranspiration into evaporation and transpiration is plotted in correspondence to leaf area per unit surface of soil below it. At sowing nearly 100% of ET comes from evaporation, while at full crop cover more than 90% of ET comes from transpiration.

Units

The evapotranspiration rate is normally expressed in millimetres (mm) per unit time. The rate expresses the amount of water lost from a cropped surface in units of water depth. The time unit can be an hour, day, decade, month or even an entire growing period or year.

As one hectare has a surface of 10000 m² and 1 mm is equal to 0.001 m, a loss of 1 mm of water corresponds to a loss of 10 m³ of water per hectare. In other words, 1 mm day⁻¹ is equivalent to 10 m³ ha⁻¹ day⁻¹.

Water depths can also be expressed in terms of energy received per unit area. The energy refers to the energy or heat required to vaporize free water. This energy, known as the latent heat of vaporization (l), is a function of the water temperature. For example, at 20° C, l is about 2.45 MJ kg⁻¹. In other words, 2.45 MJ are needed to vaporize 1 kg or

0.001 m³ of water. Hence, an energy input of 2.45 MJ per m² is able to vaporize 0.001 m or 1 mm of water, and therefore 1 mm of water is equivalent to 2.45 MJ m⁻². The evapotranspiration rate expressed in units of MJ m⁻² day⁻¹ is represented by l ET, the latent heat flux.

Table below summarizes the units used to express the evapotranspiration rate and the conversion factors.

Table: Conversion factors for evapotranspiration

	depth	volume per unit area		energy per unit area *
	mm day⁻¹	m³ ha⁻¹ day⁻¹	l s⁻¹ ha⁻¹	MJ m⁻² day⁻¹
1 mm day⁻¹	1	10	0.116	2.45
1 m³ ha⁻¹ day⁻¹	0.1	1	0.012	0.245
1 l s⁻¹ ha⁻¹	8.640	86.40	1	21.17
1 MJ m⁻² day⁻¹	0.408	4.082	0.047	1

* For water with a density of 1000 kg m⁻³ and at 20° C.

Example: Converting evaporation from one unit to another

On a summer day, net solar energy received at a lake reaches 15 MJ per square metre per day. If 80% of the energy is used to vaporize water, how large could the depth of evaporation be?			
From Table above:	1 MJ m⁻² day⁻¹ =	0.408	mm day⁻¹
Therefore:	0.8 x 15 MJ m⁻² day⁻¹ = 0.8 x 15 x 0.408 mm d⁻¹ =	4.9	mm day⁻¹
The evaporation rate could be 4.9 mm/day			

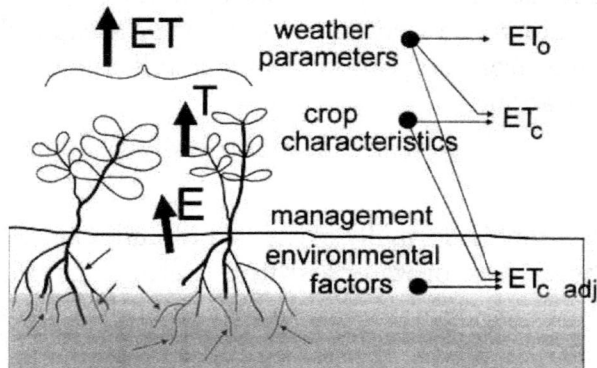

Figure: Factors affecting evapotranspiration with reference to related ET concepts.

Factors Affecting Evapotranspiration

- Weather parameters

- Crop factors

- Management and environmental conditions

Weather parameters, crop characteristics, management and environmental aspects are factors affecting evaporation and transpiration.

Weather Parameters

The principal weather parameters affecting evapotranspiration are radiation, air temperature, humidity and wind speed. Several procedures have been developed to assess the evaporation rate from these parameters. The evaporation power of the atmosphere is expressed by the reference crop evapotranspiration (ET_o). The reference crop evapotranspiration represents the evapotranspiration from a standardized vegetated surface.

Crop Factors

The crop type, variety and development stage should be considered when assessing the evapotranspiration from crops grown in large, well-managed fields. Differences in resistance to transpiration, crop height, crop roughness, reflection, ground cover and crop rooting characteristics result in different ET levels in different types of crops under identical environmental conditions. Crop evapotranspiration under standard conditions (ET_c) refers to the evaporating demand from crops that are grown in large fields under optimum soil water, excellent management and environmental conditions, and achieve full production under the given climatic conditions.

Management and Environmental Conditions

Factors such as soil salinity, poor land fertility, limited application of fertilizers, the presence of hard or impenetrable soil horizons, the absence of control of diseases and pests and poor soil management may limit the crop development and reduce the evapotranspiration. Other factors to be considered when assessing ET are ground cover, plant density and the soil water content. The effect of soil water content on ET is conditioned primarily by the magnitude of the water deficit and the type of soil. On the other hand, too much water will result in waterlogging which might damage the root and limit root water uptake by inhibiting respiration.

When assessing the ET rate, additional consideration should be given to the range of management practices that act on the climatic and crop factors affecting the ET process. Cultivation practices and the type of irrigation method can alter the microclimate, affect the crop characteristics or affect the wetting of the soil and crop surface. A windbreak reduces wind velocities and decreases the ET rate of the field directly beyond the barrier. The effect can be significant especially in windy, warm and dry conditions although evapotranspiration from the trees themselves may offset any reduction in the field. Soil evaporation in a young orchard, where trees are widely spaced, can be reduced by using a well-designed drip or trickle irrigation system. The drippers apply water directly to the soil near trees, thereby leaving the major part of the soil surface dry, and limiting the evaporation losses. The use of mulches, especially when the crop is small, is another way of substantially reducing soil evaporation. Anti-transpirants, such as stomata-closing, film-forming or reflecting material, reduce the water losses from the crop and hence the transpiration rate.

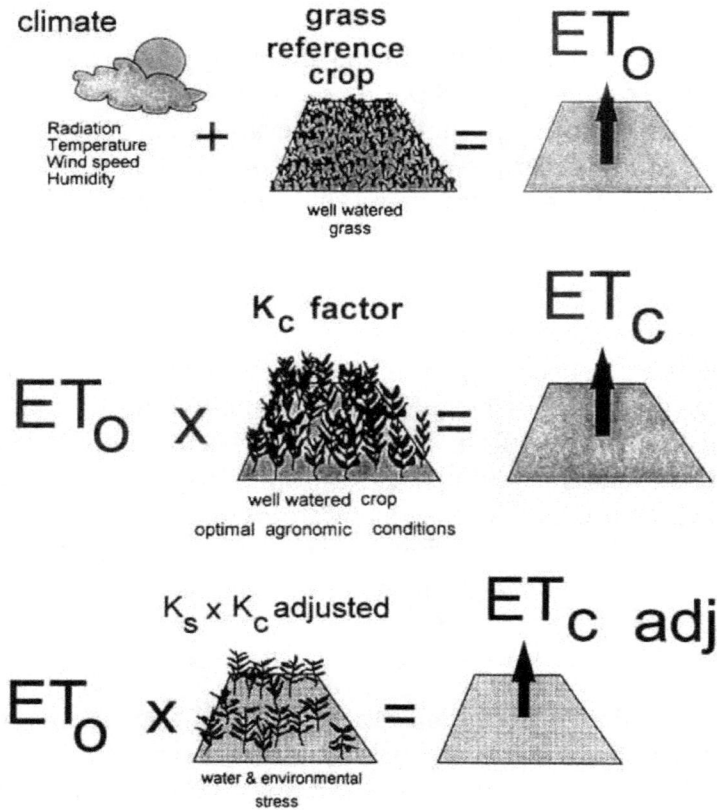

Figure: Reference (ET_o), crop evapotranspiration under standard (ET_c) and non-standard conditions ($ET_{c\,adj}$)

Where field conditions differ from the standard conditions, correction factors are required to adjust ET_c. The adjustment reflects the effect on crop evapotranspiration of the environmental and management conditions in the field.

Evapotranspiration Concepts

- Reference crop evapotranspiration (ET_o)

- Crop evapotranspiration under standard conditions (ET_c)

- Crop evapotranspiration under non-standard conditions ($ET_{c\,adj}$)

Distinctions are made between reference crop evapotranspiration (ET_o), crop evapotranspiration under standard conditions (ET_c) and crop evapotranspiration under non-standard conditions ($ET_{c\,adj}$). ET_o is a climatic parameter expressing the evaporation power of the atmosphere. ET_c refers to the evapotranspiration from excellently managed, large, well-watered fields that achieve full production under the given climatic conditions. Due to sub-optimal crop management and environmental constraints that affect crop growth and limit evapotranspiration, ET_c under non-standard conditions generally requires a correction.

Reference Crop Evapotranspiration (ET_o)

The evapotranspiration rate from a reference surface, not short of water, is called the reference crop evapotranspiration or reference evapotranspiration and is denoted as ET_o. The reference surface is a hypothetical grass reference crop with specific characteristics. The use of other denominations such as potential ET is strongly discouraged due to ambiguities in their definitions.

The concept of the reference evapotranspiration was introduced to study the evaporative demand of the atmosphere independently of crop type, crop development and management practices. As water is abundantly available at the reference evapotranspiring surface, soil factors do not affect ET. Relating ET to a specific surface provides a reference to which ET from other surfaces can be related. It obviates the need to define a separate ET level for each crop and stage of growth. ET_o values measured or calculated at different locations or in different seasons are comparable as they refer to the ET from the same reference surface.

The only factors affecting ET_o are climatic parameters. Consequently, ET_o is a climatic parameter and can be computed from weather data. ET_o expresses the evaporating power of the atmosphere at a specific location and time of the year and does not consider the crop characteristics and soil factors. The FAO Penman-Monteith method is recommended as the sole method for determining ET_o. The method has been selected because it closely approximates grass ET_o at the location evaluated, is physically based, and explicitly incorporates both physiological and aerodynamic parameters. Moreover, procedures have been developed for estimating missing climatic parameters.

Crop Evapotranspiration Under Standard Conditions (ET_c)

The crop evapotranspiration under standard conditions, denoted as ET_c, is the evapotranspiration from disease-free, well-fertilized crops, grown in large fields, under optimum soil water conditions, and achieving full production under the given climatic conditions.

Table: Average ET_o for different agroclimatic regions in mm/day

Regions	Mean daily temperature (° C)		
Cool ~10° C	Moderate 20° C	Warm > 30° C	
Tropics and subtropics			
- humid and sub-humid	2 - 3	3 - 5	5 - 7

	-arid and semi-arid	2 - 4	4 - 6	6 - 8
Temperate region				
	- humid and sub-humid	1 - 2	2 - 4	4 - 7
	-arid and semi-arid	1 - 3	4 - 7	6 - 9

The amount of water required to compensate the evapotranspiration loss from the cropped field is defined as crop water requirement. Although the values for crop evapotranspiration and crop water requirement are identical, crop water requirement refers to the amount of water that needs to be supplied, while crop evapotranspiration refers to the amount of water that is lost through evapotranspiration. The irrigation water requirement basically represents the difference between the crop water requirement and effective precipitation. The irrigation water requirement also includes additional water for leaching of salts and to compensate for non-uniformity of water application. Calculation of the irrigation water requirement is not covered in this publication, but will be the topic of a future Irrigation and Drainage Paper.

Crop evapotranspiration can be calculated from climatic data and by integrating directly the crop resistance, albedo and air resistance factors in the Penman-Monteith approach. As there is still a considerable lack of information for different crops, the Penman-Monteith method is used for the estimation of the standard reference crop to determine its evapotranspiration rate, i.e., ET_o. Experimentally determined ratios of ET_c/ET_o, called crop coefficients (K_c), are used to relate ET_c to ET_o or $ET_c = K_c\, ET_o$.

Differences in leaf anatomy, stomatal characteristics, aerodynamic properties and even albedo cause the crop evapotranspiration to differ from the reference crop evapotranspiration under the same climatic conditions. Due to variations in the crop characteristics throughout its growing season, K_c for a given crop changes from sowing till harvest.

Crop Evapotranspiration Under Non-standard Conditions ($ET_{c\,adj}$)

The crop evapotranspiration under non-standard conditions ($ET_{c\,adj}$) is the evapotranspiration from crops grown under management and environmental conditions that differ from the standard conditions. When cultivating crops in fields, the real crop evapotranspiration may deviate from ET_c due to non-optimal conditions such as the presence of pests and diseases, soil salinity, low soil fertility, water shortage or waterlogging. This may result in scanty plant growth, low plant density and may reduce the evapotranspiration rate below ET_c.

The crop evapotranspiration under non-standard conditions is calculated by using a water stress coefficient K_s and by adjusting K_c for all kinds of other stresses and environmental constraints on crop evapotranspiration.

Determining Evapotranspiration

- ET measurement

- ET computed from meteorological data

- ET estimated from pan evaporation

ET Measurement

Evapotranspiration is not easy to measure. Specific devices and accurate measurements of various physical parameters or the soil water balance in lysimeters are required to determine evapotranspiration. The methods are often expensive, demanding in terms of accuracy of measurement and can only be fully exploited by well-trained research personnel. Although the methods are inappropriate for routine measurements, they remain important for the evaluation of ET estimates obtained by more indirect methods.

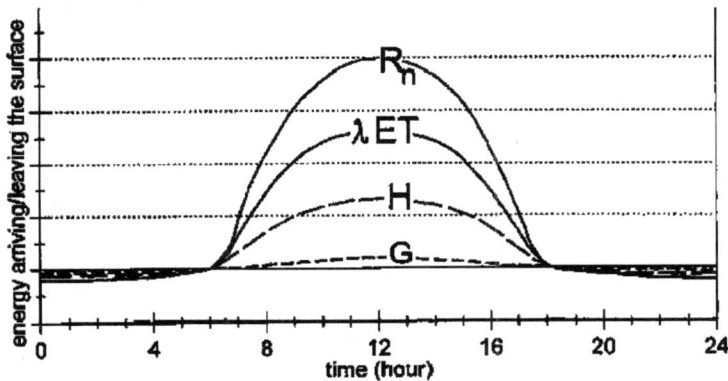

Figure: Schematic presentation of the diurnal variation of the components of the energy balance above a well-watered transpiring surface on a cloudless day

Energy Balance and Microclimatological Methods

Evaporation of water requires relatively large amounts of energy, either in the form of sensible heat or radiant energy. Therefore the evapotranspiration process is governed by energy exchange at the vegetation surface and is limited by the amount of energy available. Because of this limitation, it is possible to predict the evapotranspiration rate by applying the principle of energy conservation. The energy arriving at the surface must equal the energy leaving the surface for the same time period.

All fluxes of energy should be considered when deriving an energy balance equation. The equation for an evaporating surface can be written as:

$$R_n - G - \lambda ET - H = 0$$

where R_n is the net radiation, H the sensible heat, G the soil heat flux and λET the latent heat flux. The various terms can be either positive or negative. Positive R_n supplies energy to the surface and positive G, λ ET and H remove energy from the surface.

In Equation R_n - G - λ ET - H = 0 only vertical fluxes are considered and the net rate at which energy is being transferred horizontally, by advection, is ignored. Therefore the equation is to be applied to large, extensive surfaces of homogeneous vegetation only. The equation is restricted to the four components: R_n, λ ET, H and G. Other energy terms, such as heat stored or released in the plant, or the energy used in metabolic activities, are not considered These terms account for only a small fraction of the daily net radiation and can be considered negligible when compared with the other four components.

The latent heat flux (λ ET) representing the evapotranspiration fraction can be derived from the energy balance equation if all other components are known. Net radiation (R_n) and soil heat fluxes (G) can be measured or estimated from climatic parameters. Measurements of the sensible heat (H) are however complex and cannot be easily obtained. H requires accurate measurement of temperature gradients above the surface.

Another method of estimating evapotranspiration is the mass transfer method. This approach considers the vertical movement of small parcels of air (eddies) above a large homogeneous surface. The eddies transport material (water vapor) and energy (heat, momentum) from and towards the evaporating surface. By assuming steady state conditions and that the eddy transfer coefficients for water vapor are proportional to those for heat and momentum, the evapotranspiration rate can be computed from the vertical gradients of air temperature and water vapor via the Bowen ratio. Other direct measurement methods use gradients of wind speed and water vapor. These methods and other methods such as eddy covariance, require accurate measurement of vapor pressure, and air temperature or wind speed at different levels above the surface. Therefore, their application is restricted to primarily research situations.

Soil Water Balance

Evapotranspiration can also be determined by measuring the various components of the soil water balance. The method consists of assessing the incoming and outgoing water flux into the crop root zone over some time period. Irrigation (I) and rainfall (P) add water to the root zone. Part of I and P might be lost by surface runoff (RO) and by deep percolation (DP) that will eventually recharge the water table. Water might also be transported upward by capillary rise (CR) from a shallow water table towards the root zone or even transferred horizontally by subsurface flow in (SF_{in}) or out of (SF_{out}) the root zone. In many situations, however, except under conditions with large slopes, SF_{in} and SF_{out} are minor and can be ignored. Soil evaporation and crop transpiration deplete water from the root zone. If all fluxes other than evapotranspiration (ET) can be assessed, the evapotranspiration can be deduced from the change in soil water content (Δ SW) over the time period:

$$ET = I + P - RO - DP + CR \pm \Delta SF \pm \Delta SW$$

Some fluxes such as subsurface flow, deep percolation and capillary rise from a water table are difficult to assess and short time periods cannot be considered. The soil water balance method can usually only give ET estimates over long time periods of the order of week-long or ten-day periods.

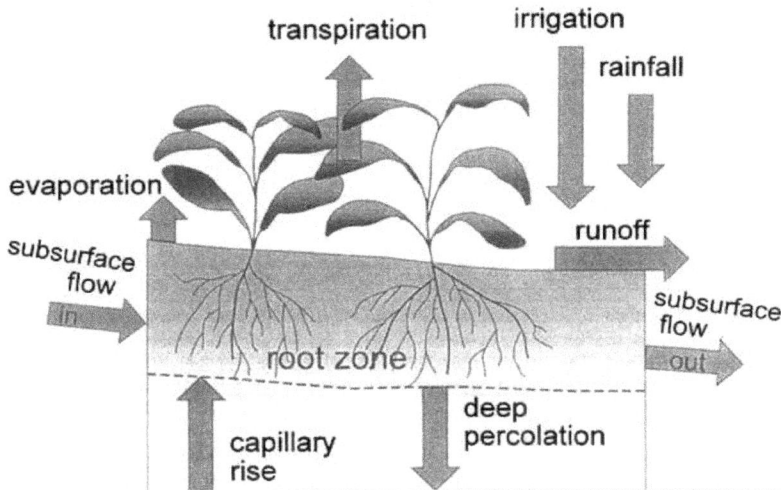

Figure: Soil water balance of the root zone.

Lysimeters

By isolating the crop root zone from its environment and controlling the processes that are difficult to measure, the different terms in the soil water balance equation can be determined with greater accuracy. This is done in lysimeters where the crop grows in isolated tanks filled with either disturbed or undisturbed soil. In precision weighing lysimeters, where the water loss is directly measured by the change of mass, evapotranspiration can be obtained with an accuracy of a few hundredths of a millimetre, and small time periods such as an hour can be considered. In non-weighing lysimeters the evapotranspiration for a given time period is determined by deducting the drainage water, collected at the bottom of the lysimeters, from the total water input.

A requirement of lysimeters is that the vegetation both inside and immediately outside of the lysimeter be perfectly matched (same height and leaf area index). This requirement has historically not been closely adhered to in a majority of lysimeter studies and has resulted in severely erroneous and unrepresentative ET_c and K_c data.

As lysimeters are difficult and expensive to construct and as their operation and maintenance require special care, their use is limited to specific research purposes.

ET Computed from Meteorological Data

Owing to the difficulty of obtaining accurate field measurements, ET is commonly computed from weather data. A large number of empirical or semi-empirical equations have been developed for assessing crop or reference crop evapotranspiration from meteorological data. Some of the methods are only valid under specific climatic and agronomic conditions and cannot be applied under conditions different from those under which they were originally developed.

Numerous researchers have analysed the performance of the various calculation methods for different locations. As a result of an Expert Consultation held in May 1990, the FAO Penman-Monteith method is now recommended as the standard method for the definition and computation of the reference evapotranspiration, ET_o. The ET from crop surfaces under standard conditions is determined by crop coefficients (K_c) that relate ET_c to ET_o. The ET from crop surfaces under non-standard conditions is adjusted by a water stress coefficient (K_s) or by modifying the crop coefficient.

ET Estimated from Pan Evaporation

Evaporation from an open water surface provides an index of the integrated effect of radiation, air temperature, air humidity and wind on evapotranspiration. However, differences in the water and cropped surface produce significant differences in the water loss from an open water surface and the crop. The pan has proved its practical value and has been used successfully to estimate reference evapotranspiration by observing the evaporation loss from a water surface and applying empirical coefficients to relate pan evaporation to ET_o.

Condensation

Clouds over Kiger Notch, Steen's Mountain, Oregon

Condensation is the process by which water vapor in the air is changed into liquid water. Condensation is crucial to the water cycle because it is responsible for the formation of clouds. These clouds may produce precipitation, which is the primary route for water to return to the Earth's surface within the water cycle. Condensation is the opposite of evaporation.

You don't have to look at something as far away as a cloud to notice condensation, though. Condensation is responsible for ground-level fog, for your glasses fogging up when you go from a cold room to the outdoors on a hot, humid day, for the water that drips off the outside of your glass of iced tea, and for the water on the inside of the windows in your home on a cold day.

The phase change that accompanies water as it moves between its vapor, liquid, and solid form is exhibited in the arrangement of water molecules. Water molecules in the vapor form are arranged more randomly than in liquid water. As condensation occurs and liquid water forms from the vapor, the water molecules become organized in a less random structure, which is less random than in vapor, and heat is released into the atmosphere as a result.

Condensation in the Air

Even though clouds are absent in a crystal clear blue sky, water is still present in the form of water vapor and droplets which are too small to be seen. Depending on weather conditions, water molecules will combine with tiny particles of dust, salt, and smoke in the air to form cloud droplets, which grow and develop into clouds, a form of water we can see. Cloud droplets can vary greatly in size, from 10 microns to 1 millimeter, and even as large as 5 mm. This process occurs higher in the sky where the air is cooler and more condensation occurs relative to evaporation. As water droplets combine with each other, and grow in size, clouds not only develop, but precipitation may also occur. Precipitation is essentially water in its liquid or solid form falling from the base of a cloud. This seems to happen too often during picnics or when large groups of people gather at swimming pools.

Reasons for Lower Temperatures at Higher Altitudes

The clouds formed by condensation are an intricate and critical component of Earth's environment. Clouds regulate the flow of radiant energy into and out of Earth's climate system. They influence the Earth's climate by reflecting incoming solar radiation back to space and outgoing radiation from the Earth's surface. Often at night, clouds act as a "blanket," keeping a portion of the day's heat next to the surface. Changing cloud patterns modify the Earth's energy balance, and, in turn, temperatures on the Earth's surface.

Cumulonimbus cloud

As we said, clouds form in the atmosphere because air containing water vapor rises and cools. The key to this process is that air near the Earth's surface is warmed by solar radiation. But, do you know why the atmosphere cools above the Earth's surface? Generally, air pressure, is the reason. Air has mass and at sea level the weight of a column of air pressing down on your head is about 14 ½ pounds per square inch. The pressure, called barometric pressure, that results is a consequence of the density of the air above. At higher altitudes, there is less air above, and, thus, less air pressure pressing down. The barometric pressure is lower, and lower barometric pressure is associated with fewer molecules per unit volume. Therefore, the air at higher altitudes is less dense. As the total heat content of a system is directly related to the amount of matter present, it is cooler at higher elevation because fewer air molecules exist in a certain volume of air higher up. This means cooler air.

Condensation Near the Ground

Condensation also occurs at ground level, as this picture of a cloud bank in California shows. The difference between fog and clouds which form above the Earth's surface is that rising air is not required to form fog. Fog develops when air having a relatively high humidity comes in contact with a colder surface, often the Earth's surface, and cools to the dew point. Additional cooling leads to condensation and the growth of low-level clouds. Fog that develops when warmer air moves over a colder surface is known as advective fog. Another form of fog, known as radiative fog, develops at night when surface temperatures cool. If the air is still, the fog layer does not readily mix with the air above it, which encourages the development of shallow ground fog.

Condensation on your Glass

Picture of glasses fogged with condensation after being chilled and going into a warm, moist room. You probably see condensation right at home every day. If you wear glasses and go from a cold, air-conditioned room to outside on a humid day, the lenses fog up as small water droplets coat the surface via condensation. People buy coasters to keep condensed water from dripping off their chilled drink glass onto their coffee tables. Condensation is responsible for the water covering the inside of a window on a cold day and for the moisture on the inside of car windows, especially after people have been exhaling moist air. All of these are examples of water leaving the vapor state in the warm air and condensing into liquid as it is cools.

Reasons of Cloud Formation and Rainfall

Air, even "clear air," contains water molecules. Clouds exist in the atmosphere because of rising air. As air rises and cools the water in it can "condense out", forming clouds. Since clouds drift over the landscape, they are one of the ways that water moves geographically around the globe in the water cycle. A common myth is that clouds form because cooler air can hold less water than warmer air—but this is not true.

As Alistair Fraser explains, "What appears to be cloud-free air (virtually) always contains sub microscopic drops, but as evaporation exceeds condensation, the drops do not survive long after an initial chance clumping of molecules. As air is cooled, the evaporation rate decreases more rapidly than does the condensation rate with the result that there comes a temperature where the evaporation is less than the condensation and a droplet can grow into a cloud drop. When the temperature drops below the dew-point temperature, there is a net condensation and a cloud forms,".

Contrails: Man-made Clouds

Condensation trails made by a high-flying airplanes, over Lake Jackson, Florida.

You've seen the cloud-like trails that high-flying airplanes leave behind and you probably know they are called contrails. Maybe you didn't know they were called that because they are actually condensation trails and, in fact, are not much different than natural clouds. If the exhaust from the airplane contains water vapor, and if the air is very cold, then the water vapor in the exhaust will condense out into what is essentially a cirrus cloud.

As a matter of fact, sailors have known for some time to look specifically at the patterns and persistence of jet contrails for weather forecasting. On days where the contrails disappear quickly or don't even form, they can expect continuing good weather, while on days where they persist, a change in the weather pattern may be expected. Contrails are a concern in climate studies as increased jet traffic may result in an increase in cloud cover. Several scientific studies are being conducted with respect to contrail formation and their impact on climates. Cirrus clouds affect Earth's climate by reflecting incoming sunlight and inhibiting heat loss from the surface of the planet. It has been estimated that in certain heavy air-traffic corridors, cloud cover has increased by as much as 20 percent.

Precipitation

Precipitation is one of the three main processes (evaporation, condensation, and precipitation) that constitute the hydrologic cycle, the continual exchange of water between the atmosphere and Earth's surface. Water evaporates from ocean, land, and freshwater surfaces, is carried aloft as vapour by the air currents, condenses to form clouds, and ultimately is returned to Earth's surface as precipitation. The average global stock of water vapour in the atmosphere is equivalent to a layer of water 2.5 cm (1 inch) deep covering the whole Earth. Because Earth's average annual rainfall is about 100 cm (39 inches), the average time that the water spends in the atmosphere, between its evaporation from the surface and its return as precipitation, is about 1/40 of a year, or about nine days. Of the water vapour that is carried at all heights across a given region by the winds, only a small percentage is converted into precipitation and reaches the ground in that area. In deep and extensive cloud systems, the conversion is more efficient, but even in thunderclouds the quantities of rain and hail released amount to only some 10 percent of the total moisture entering the storm.

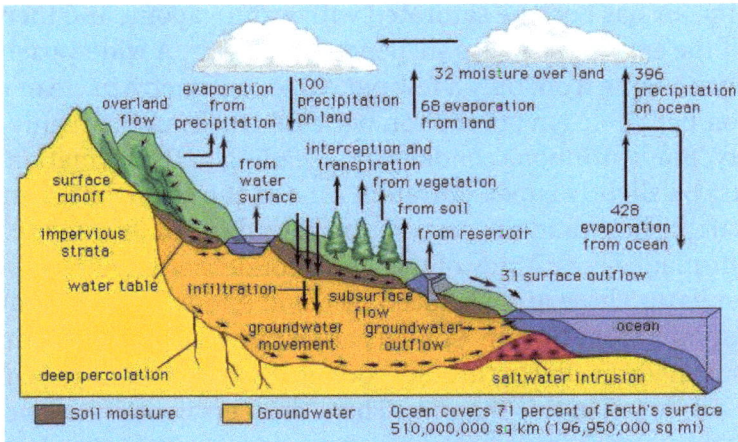

In the hydrologic cycle, water is transferred between the land surface, the ocean, and the atmosphere. The numbers on the arrows indicate relative water fluxes.

In the measurement of precipitation, it is necessary to distinguish between the amount—defined as the depth of precipitation (calculated as though it were all rain) that has fallen at a given point during a specified interval of time—and the rate or intensity, which specifies the depth of water that has fallen at a point during a particular interval of time. Persistent moderate rain, for example, might fall at an average rate of 5 mm per hour (0.2 inch per hour) and thus produce 120 mm (4.7 inches) of rain in 24 hours. A thunderstorm might produce this total quantity of rain in 20 minutes, but at its peak intensity the rate of rainfall might become much greater—perhaps 120 mm per hour (4.7 inches per hour), or 2mm (0.08 inch) per minute—for a minute or two.

The amount of precipitation falling during a fixed period is measured regularly at many thousands of places on Earth's surface by rather simple rain gauges. Measurement of precipitation intensity requires a recording rain gauge, in which water falling into a collector of known surface area is continuously recorded on a moving chart or a magnetic tape. Investigations are being carried out on the feasibility of obtaining continuous measurements of rainfall over large catchment areas by means of radar.

Apart from the trifling contributions made by dew, frost, and rime, as well as desalination plants, the sole source of fresh water for sustaining rivers, lakes, and all life on Earth is provided by precipitation from clouds. Precipitation is therefore indispensable and overwhelmingly beneficial to humankind, but extremely heavy rainfall can cause great harm: soil erosion, landslides, and flooding. Hailstorm damage to crops, buildings, and livestock can prove very costly.

Origin of Precipitation in Clouds

Cloud Formation

Clouds are formed by the lifting of damp air, which cools by expansion as it encounters the lower pressures existing at higher levels in the atmosphere. The relative humidity increases until the air has become saturated with water vapour, and then condensation occurs on any of the aerosol particles suspended in the air. A wide variety of these exist in concentrations ranging from only a few per cubic centimetre in clean maritime air to perhaps 1 million per cubic cm (16 million per cubic inch) in the highly polluted air of an industrial city. For continuous condensation leading to the formation of cloud droplets, the air must be slightly supersaturated. Among the highly efficient condensation nuclei are sea-salt particles and the particles produced by combustion (e.g., natural forest fires and man-made fires). Many of the larger condensation nuclei over land consist of ammonium sulfate. These are produced by cloud and fog droplets absorbing sulfur dioxide and ammonia from the air. Condensation onto the nuclei continues as rapidly as water vapour is made available through cooling; droplets about 10 μm (0.0004 inch) in diameter are produced in this manner. These droplets constitute a nonprecipitating cloud.

Types of Precipitation

Drizzle

Liquid precipitation in the form of very small drops, with diameters between 0.2 and 0.5 mm (0.008 and 0.02 inch) and terminal velocities between 70 and 200 cm per second (28 and 79 inches per second), is defined as drizzle. It forms by the coalescence of even smaller droplets in low-layer clouds containing weak updrafts of only a few centimetres per second. High relative humidity below the cloud base is required to prevent the drops from evaporating before reaching the ground; drizzle is classified as slight, moderate, or thick. Slight drizzle produces negligible runoff from the roofs of buildings, and thick drizzle accumulates at a rate in excess of 1 mm per hour (0.04 inch per hour).

Rain and Freezing Rain

Liquid waterdrops with diameters greater than those of drizzle constitute rain. Raindrops rarely exceed 6 mm (0.2 inch) in diameter because they become unstable when larger than this and break up during their fall. The terminal velocities of raindrops at ground level range from 2 metres per second (7 feet per second) for the smallest to about 10 metres per second (30 feet per second) for the largest. The smaller raindrops are kept nearly spherical by surface-tension forces, but, as the diameter surpasses about 2 mm (0.08 inch), they become increasingly flattened by aerodynamic forces. When the diameter reaches 6 mm, the undersurface of the drop becomes concave because of the airstream, and the surface of the drop is sheared off to form a rapidly expanding "bubble" or "bag" attached to an annular ring containing the bulk of the water. Eventually the bag bursts into a spray of fine droplets, and the ring breaks up into a circlet of millimetre-sized drops.

A rain shaft piercing a tropical sunset as seen from Man-o'-War Bay, Tobago, Caribbean Sea.

Rain of a given intensity is composed of a spectrum of drop sizes, the average and median drop diameters being larger in rains of greater intensity. The largest drops, which have a diameter greater than 5 mm (0.2 inch), appear only in the heavy rains of intense storms.

When raindrops fall through a cold layer of air (colder than 0° C, or 32° F) and become supercooled, freezing rain occurs. The drops may freeze on impact with the ground to form a very slippery and dangerous "glazed" ice that is difficult to see because it is almost transparent.

Snow and Sleet

Snow in the atmosphere can be subdivided into ice crystals and snowflakes. Ice crystals generally form on ice nuclei at temperatures appreciably below the freezing point. Below −40 °C (−40 °F) water vapour can solidify without the presence of a nucleus. Snowflakes are aggregates of ice crystals that appear in an infinite variety of shapes, mainly at temperatures near the freezing point of water.

In British terminology, sleet is the term used to describe precipitation of snow and rain together or of snow melting as it falls. In the United States, it is used to denote partly frozen ice pellets.

Colorado's fine, light snow attracts millions of skiers every year.

Snow crystals generally have a hexagonal pattern, often with beautifully intricate shapes. Three- and 12-branched forms occur occasionally. The hexagonal form of the atmospheric ice crystals, their varying size and shape notwithstanding, is an outward manifestation of an internal arrangement in which the oxygen atoms form an open lattice (network) with hexagonally symmetrical structure. According to a recent internationally accepted classification, there are seven types of snow crystals: plates, stellars, columns, needles, spatial dendrites, capped columns, and irregular crystals. The size and shape of the snow crystals depend mainly on the temperature of their formation and on the amount of water vapour that is available for deposition. The two principal influences are not independent; the possible water vapour admixture of the air decreases strongly with decreasing temperature. The vapour pressure in equilibrium, or state of balance, with a level surface of pure ice is 50 times greater at $-2°$ C ($28°$ F) than at $-42°$ C ($-44°$ F), the likely limits of snow formation in the air. Crystal shape and temperature at formation are related in the table.

Ice crystal shape and temperature at formation	
temperature (degrees Celsius)	form
0 to −3	thin hexagonal plates
−3 to −5	needles
−5 to −8	hollow, prismatic columns
−8 to −12	hexagonal plates
−12 to −16	dendritic crystals
−16 to −25	hexagonal plates
−25 to −50	hollow prisms

At temperatures above about $-40°$ C, the crystals form on nuclei of very small size that float in the air (heterogeneous nucleation). The nuclei consist predominantly of silicate minerals of terrestrial origin, mainly clay minerals and micas. At still lower temperatures, ice may form directly from water vapour (homogeneous nucleation). The influence of the atmospheric water vapour depends mainly on its degree of supersaturation with respect to ice.

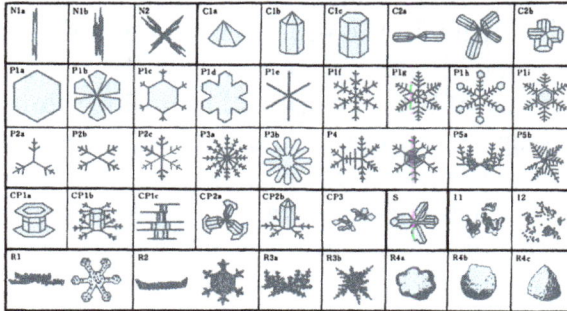

Figure: Classification of frozen precipitation.

If the air contains a large excess of water vapour, the snow particles will grow fast, and there may be a tendency for dendritic (branching) growth. With low temperature, the excess water vapour tends to be small, and the crystals remain small. In relatively dry layers, the snow particles generally have simple forms. Complicated forms of crystals will cling together with others to form snowflakes that consist occasionally of up to 100 crystals; the diameter of such flakes may be as large as 2.5 cm (1 inch). This process will be furthered if the crystals are near the freezing point and wet, possibly by collision with undercooled water droplets. If a crystal falls into a cloud with great numbers of such drops, it will sweep some of them up. Coming into contact with ice, they freeze and form an ice cover around the crystal. Such particles are called soft hail or graupel .

Snow particles constitute the clouds of cirrus type—namely cirrus, cirrostratus, and cirrocumulus—and many clouds of alto type. Ice and snow clouds originate normally only at temperatures some degrees below the freezing point; they predominate at –20° C (–4° F). In temperate and low latitudes these clouds occur in the higher layers of the troposphere. In tropical regions they hardly ever occur below 4,570 metres (15,000 feet). On high mountains and particularly in polar regions, they can occur near the surface and may appear as ice fogs. If cold air near the ground is overlain by warmer air (a very common occurrence in polar regions, especially in winter), mixture at the border leads to supersaturation in the cold air. Small ice columns and needles, "diamond dust," will be formed and will float down, glittering, even from a cloudless sky. In the coldest parts of Antarctica, where temperatures near the surface are below –50° C (–58° F) on the average and rarely above –30° C (–22° F), the formation of diamond dust is a common occurrence. The floating and falling ice crystals produce in the light of the Sun and the Moon the manifold phenomena of atmospheric optics, halos, arcs, circles, mock suns, some coronas, and iridescent clouds. Most of the different optical appearances can be explained by the shapes of the crystals and their position with respect to the light source.

Most of the moderate to heavy rain in temperate latitudes depends on the presence of ice and snow particles in clouds. In the free atmosphere, droplets of fluid water can be undercooled considerably; typical ice clouds originate mainly at a temperature near −20° C. At an identical temperature below the freezing point, the water molecules are kept more firmly in the solid than in the fluid state. The equilibrium pressure of the gaseous phase is smaller in contact with ice than with water. At −20° C, which is the temperature of the formation of typical ice clouds (cirrus), the equilibrium pressure with respect to undercooled water (relative humidity 100 percent) is 22 percent greater than the equilibrium pressure of the water vapour in contact with ice. Hence, with an excess of water vapour beyond the equilibrium state, the ice particles tend to incorporate more water vapour and to grow more rapidly than the water droplets.

Being larger and so less retarded by friction, the ice particles fall more rapidly. In their fall they sweep up some water droplets, which on contact become frozen. Thus, a cloud layer originally consisting mainly of undercooled water with few ice crystals is transformed into an ice cloud. The development of the anvil shape at the top of a towering cumulonimbus cloud shows this transformation very clearly. The larger ice particles overcome more readily the rising tendency of the air in the cloud. Falling into lower levels they grow, aggregating with other crystals and possibly with waterdrops, melt, and form raindrops when near-surface temperatures permit.

Hail

Solid precipitation in the form of hard pellets of ice that fall from cumulonimbus clouds is called hail. It is convenient to distinguish between three types of hail particles.

The first is soft hail, or snow pellets, which are white opaque rounded or conical pellets as large as 6 mm (0.2 inch) in diameter. They are composed of small cloud droplets frozen together, have a low density, and are readily crushed.

The second is small hail (ice grains or pellets), which are transparent or translucent pellets of ice that are spherical, spheroidal, conical, or irregular in shape, with diameters of a few millimetres. They may consist of frozen raindrops, of largely melted and refrozen snowflakes, or of snow pellets encased in a thin layer of solid ice.

True hailstones, the third type, are hard pellets of ice, larger than 5 mm (0.2 inch) in diameter, that may be spherical, spheroidal, conical, discoidal, or irregular in shape and often have a structure of concentric layers of alternately clear and opaque ice. A moderately severe storm may produce stones a few centimetres in diameter, whereas a very severe storm may release stones with a maximum diameter of 10 cm (4 inches) or more. Large damaging hail falls most frequently in the continental areas of middle latitudes (e.g., in the Nebraska-Wyoming-Colorado area of the United States, in South Africa, and in northern India) but is rare in equatorial regions. Terminal velocities of hailstones range from about 5 metres (16 feet) per second for the smallest stones to perhaps 40 metres (130 feet) per second for stones 5 cm (2 inches) in diameter.

World Distribution of Precipitation

Regional and Latitudinal Distribution

The yearly precipitation averaged over the whole Earth is about 100 cm (39 inches), but this is distributed very unevenly. The regions of highest rainfall are found in the equatorial zone and the monsoon area of Southeast Asia. Middle latitudes receive moderate amounts of precipitation, but little falls in the desert regions of the subtropics and around the poles.

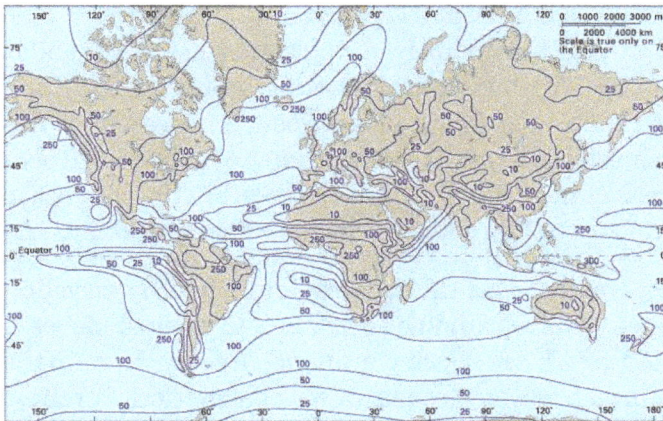

Global distribution of mean annual rainfall (in centimetres).

If Earth's surface were perfectly uniform, the long-term average rainfall would be distributed in distinct latitudinal bands, but the situation is complicated by the pattern of the global winds, the distribution of land and sea, and the presence of mountains. Because rainfall results from the ascent and cooling of moist air, the areas of heavy rain indicate regions of rising air, whereas the deserts occur in regions in which the air is warmed and dried during descent. In the subtropics, the trade winds bring plentiful rain to the east coasts of the continents, but the west coasts tend to be dry. On the other hand, in high latitudes the west coasts are generally wetter than the east coasts. Rain tends to be abundant on the windward slopes of mountain ranges but sparse on the lee sides.

In the equatorial belt, the trade winds from both hemispheres converge and give rise to a general upward motion of air, which becomes intensified locally in tropical storms that produce very heavy rains in the Caribbean, the Indian and southwest Pacific oceans, and the China Sea and in thunderstorms that are especially frequent and active over the land areas. During the annual cycle, the doldrums move toward the summer hemisphere, so outside a central region near the Equator, which has abundant rain at all seasons, there is a zone that receives much rain in summer but a good deal less in winter.

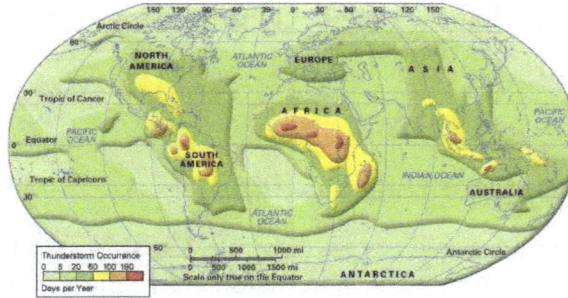

World patterns of thunderstorm frequencyThunderstorms occur most often in the tropical latitudes over land, where the air is most likely to heat quickly and form strong updrafts.

The dry areas of the subtropics—such as the desert regions of North Africa, the Arabian Peninsula, South Africa, Australia, and central South America—are due to the presence of semipermanent subtropical anticyclones in which the air subsides and becomes warm and dry. These high-pressure belts tend to migrate with the seasons and cause summer dryness on the poleward side and winter dryness on the equatorward side of their mean positions. The easterly trade winds, having made a long passage over the warm oceans, bring plentiful rains to the east coasts of the subtropical landmasses, but the west coasts and the interiors of the continents, which are often sheltered by mountain ranges, are very dry.

In middle latitudes, weather and rainfall are dominated by traveling depressions and fronts that yield a good deal of rain in all seasons and in most places except the far interiors of the Asian and North American continents. Generally, rainfall is more abundant in summer, except on the western coasts of North America, Europe, and North Africa, where it is higher during the winter.

At high latitudes and especially in the polar regions, the low precipitation is caused partly by subsidence of air in the high-pressure belts and partly by the low temperatures. Snow or rain occur at times, but evaporation from the cold sea and land surfaces is slow, and the cold air has little capacity for moisture.

The influence of oceans and continents on rainfall is particularly striking in the case of the Indian monsoon. During the Northern Hemisphere winter, cool dry air from the interior of the continent flows southward and produces little rain over the land areas. After the air has traveled some distance over the warm tropical ocean, however, it releases heavy shower rains over the East Indies. During the northern summer, when the monsoon blows from the southwest, rainfall is heavy over India and Southeast Asia. These rains are intensified where the air is forced to ascend over the windward slopes of the Western Ghats and the Himalayas.

The combined effects of land, sea, mountains, and prevailing winds show up in South America. There the desert in southern Argentina is sheltered by the Andes from the westerly winds blowing from the Pacific Ocean, and the west-coast desert not only is situated under the South Pacific subtropical anticyclone but is also protected by the Andes against rain-bearing winds from the Atlantic.

Amounts and Variability

The long-term average amounts of precipitation for a season or a year give little information on the regularity with which rain may be expected, particularly for regions where the average amounts are small. For example, at Iquique, a city in northern Chile, four years once passed without rain, whereas the fifth year gave 15 mm (0.6 inch); the five-year average was therefore 3 mm (0.1 inch). Clearly, such averages are of little practical value, and the frequency distribution or the variability of precipitation also must be known.

The variability of the annual rainfall is closely related to the average amounts. For example, over the British Isles, which have a very dependable rainfall, the annual amount varies by less than 10 percent above the long-term average value. A variability of less than 15 percent is typical of the mid-latitude cyclonic belts of the Pacific and Atlantic oceans and of much of the wet equatorial regions. In the interiors of the desert areas of Africa, Arabia, and Central Asia, however, the rainfall in a particular year may deviate from the normal long-term average by more than 40 percent. The variability for individual seasons or months may differ considerably from that for the year as a whole, but again the variability tends to be higher where the average amounts are low.

The heaviest annual rainfall in the world was recorded at the village of Cherrapunji, India, where 26,470 mm (1,042 inches) fell between August 1860 and July 1861. The heaviest rainfall in a period of 24 hours was 1,870 mm (74 inches), recorded at the village of Cilaos, Réunion, in the Indian Ocean on March 15–16, 1952. The lowest recorded rainfall in the world occurred at Arica, a port city in northern Chile. An annual average, taken over a 43-year period, was only 0.5 mm (0.02 inch).

Although past records give some guide, it is not possible to estimate very precisely the maximum possible precipitation that may fall in a given locality during a specified interval of time. Much will depend on a favourable combination of several factors, including the properties of the storm and the effects of local topography. Thus, it is possible only to make estimates that are based on analyses of past storms or on theoretical calculations that attempt to maximize statistically the various factors or the most effective combination of factors that are known to control the duration and intensity of the precipitation. For many important planning and design problems, however, estimates of the greatest precipitation to be expected at a given location within a specified number of years are required.

In the designing of a dam, the highest 24-hour rainfall to be expected once in 30 years over the whole catchment area might be relevant. For dealing with such problems, a great deal of work has been devoted to determining from past records the frequency with which rainfalls of given intensity and total amount may be expected to reoccur at particular locations and also to determining the statistics of rainfall for a specific area from measurements made at only a few points.

Effects of Precipitation

Raindrop Impact and Soil Erosion

Large raindrops, up to 6 mm (0.2 inch) in diameter, have terminal velocities of about 10 metres (30 feet) per second and so may cause considerable compaction and erosion of the soil by their force of impact. The formation of a compacted crust makes it more difficult for air and water to reach the roots of plants and encourages the water to run off the surface and carry away the topsoil with it. In hilly and mountainous areas, heavy rain may turn the soil into mud and slurry, which may produce enormous erosion by mudflow generation. Rainwater running off hard impervious surfaces or waterlogged soil may cause local flooding.

Surface Runoff

The rainwater that is not evaporated or stored in the soil eventually runs off the surface and finds its way into rivers, streams, and lakes or percolates through the rocks and becomes stored in natural underground reservoirs. A given catchment area must achieve an overall balance such that precipitation (P) less evaporation of moisture from the surface (E) will equal storage in the ground (S) and runoff (R). This may be expressed: $P - E = S + R$. The runoff may be determined by measuring the flow of water in the rivers with stream gauges, and the precipitation may be measured by a network of rain gauges, but storage and evaporation are more difficult to estimate.

Of all the water that falls on Earth's surface, the relative amounts that run off, evaporate, or seep into the ground vary so much for different areas that no firm figures can be given for Earth as a whole. It has been estimated, however, that in the United States 10 to 50 percent of the rainfall runs off at once, 10 to 30 percent evaporates, and 40 to 60 percent is absorbed by the soil. Of the entire rainfall, 15 to 30 percent is estimated to be used by plants, either to form plant tissue or in transpiration.

References

- Water-cycle condensation: usgs.gov, Retrieved 19 July 2018

- Different-steps-of-the-hydrologic-cycle: conserve-energy-future.com, Retrieved 11 July 2018

- Precipitation, climate-meteorology, science: britannica.com, Retrieved 31 March 2018

- Types-of-precipitation, climate-meteorology, science: britannica.com, Retrieved 19 May 2018

- World-distribution-of-precipitation, climate-meteorology, science: britannica.com, Retrieved 22 June 2018

6

Water Pollution

The contamination of water bodies, such as lakes, rivers, aquifers, oceans and ground-water, due to anthropogenic activities is termed as water pollution. The topics included in this chapter such as effluent, acidification of water, ocean deoxygenation, etc. addresses the important facets of water pollution.

Water has a great self-generating capacity that can neutralize the polluting interventions carried out by humans. However, if human activities continue this uncontrolled and unsustainable exploitation of this resource, this regenerating capacity shall fail and it will be jeopardized definitively.

Man is now aware of this and is increasingly aware of the mechanisms that regulate the water cycle, and allow rivers, lakes, seas and oceans to live, and know where and how to intervene.

Facts

Water pollution is intended as water quality degradation caused by the introduction of substances that alter its physicochemical characteristics and impede its normal use. These substances having either a solid, liquid or gaseous origin have different effects according to their amount and potential danger and fragility of the environments where they are released. These substances can have a human origin, when they are introduced by humans, or a natural origin. Pollution of water of natural origin can be caused by decomposition of organic debris, saltwater invading coastal aquifers; water blackening due to landslides, earthquakes, dust erupted from volcanoes. Pollution can either be found at the level of superficial water or subterranean water. Pollution of superficial water can differ in its features or seriousness depending on whether it affects water in a river or in a lake and the consequences are: fish fauna depletion, death of aerobic bacteria and aquatic plants, formation of pestilential odors and deteriorating material, diffusion of pathogenic microorganisms, moreover, the more the water is polluted the greater are the costs of making it drinkable.

Pollution of aquifers is very dangerous from the point of view of preservation of water as a resource for humans. Once it's polluted, infact, subterranean water has a low depurative power and shows to employ much more time to recover original quality of water.

Substances Polluting Water

Water used in the agricultural, industrial and civil sectors often contains substances, which will alter the ecosystem and hence must not be discharged directly into river flows.

The most common polluting agents are the following:

- Fecal pollutants: Materials of fecal origin that reach water bodies through sewage discharges or introduction of zootechnical manure that hasn't been adequately treated. If there is a high fecal type pollution, can be observed the presence of pathogen microorganisms in water that can cause diseases as cholera, typhus fever, viral hepatitis, etc.

- Toxic inorganic substances: These are constituted by heavy metal ions that can poison or kill living organisms. Industries that employ these heavy metals during processing must sanitize them to eliminate any heavy metal leftover before discharging water.

- Inorganic harmful substances: There are substances constituted by phosphates and polyphosphates existing in detergents, fertilizers, compounds of nitrogen and phosphorus and in some industrial discharges. These substances cause eutrophication.

- Unnatural organic substances: In this category are included weed killers, pesticides, insecticides, etc. These substances are convenient for agriculture but can pollute both water and soil. Among these substances are included also organic solvents used by industries, such as trichlorethylene, acetone, benzene, etc., which must be eliminated before water is discharged.

- Free oils and emulsifiers: These are insoluble, low density substances which for this reason form superficial oily film layers that prevent oxygen dissolution in water. It's not a rare phenomenon and it can cause real ecological disasters.

- Suspended solids: Creating a mixture of various kinds of substances that makes water murky and prevent solar light from passing through. When they deposit deep on the bottom of a water body they obstruct vegetation growth.

- Heat, acids and strong bases: Originating especially from industrial discharges. They reduce the solubility of oxygen; modify the temperature and pH of the environment causing pathological alterations or the disappearance of living organisms or on the contrary the comparison of others.

Alterations

The different types of pollution lead to chemical or physical alterations of the water, with mechanisms, which are very complex at times. Contaminants that are dispersed in the water have harmful effects on animal populations and plants that can be classified into the following categories:

- Deoxygenation or oxygen: Depleting effects: these effects are caused by the organic substances that are present in industrial waste. When industrial waste is dispersed in the water, it is degraded by microorganisms, with an excessive consumption of the oxygen dissolved in the water. The reduced availability of oxygen in the water is the cause of death of animal and plant species that are unable to tolerate this lack of oxygen and, on the other hand, an invasion of those species that are not influenced by the lack of oxygen;

- Eutrophication effects: Eutrophication is a phenomenon provoked by compounds (mineral nitrogen and phosphorus) that favor an abnormal growth of populations of phytoplankton and algae, at the expense of the survival of other plant and animal species. Lakes and sea-coasts are particularly affected by this;

- Physical effects: Are caused by very high temperature waste waters;

- Effects caused by radiations;

- Pathogenic and toxic effects: Are caused by wastewaters with a high content of suspended toxic or pathogenic materials (heavy metals, mineral oils, hydrocarbons, ammonia, solvents, detergents, pesticides, etc.)

Water Regenerates itself

When polluted, fresh water basins have the capacity to self-clean their water, i.e. to make the water return to its original quality and purity. This self-cleaning phenomenon is provoked by bacteria that, in the presence of oxygen, degrade and transform the polluting substances into inert inorganic compounds. Obviously this process is not effective on all types of pollutants and for any amount of polluting substances present in the water. In some cases, human intervention is necessary to clean the water reserves that have been polluted.

The type of treatment that must be used to clean the water depends essentially on the type of pollutant that must be eliminated. The greater the number of pollutants in the water, the more difficult it is to clean it. In some cases polluted water is extracted, purified and then returned to the water table, river or lake. In other cases water is purified on site (i.e. without removing the water from its natural site).

Also the sea has a great self-regenerating capacity, which can neutralize human polluting actions. However, if human activities continue to persevere in an uncontrolled and unsustainable exploitation of a resource that seems (but is not) inexhaustible, this regenerating capacity will cease and it will permanently jeopardize the sea's capacity to correctly carry out all the vital functions that it provides today.

Man is now aware of this and is increasingly aware of the mechanisms that regulate the water cycle, that allow rivers, lakes, seas and oceans to live. Therefore we know where and how to intervene: the problem is to succeed in improving and spreading good practices and sustainable use of the water resource all over the world, in our communities and in those countries where respect for natural resources seems to be applicable only to the richer countries.

DDT

The history of DDT represents a significant example of the risks for humans every time they intervene on the environment without knowing inside out the balances of ecosystems. The insecticidal power of DDT was discovered in 1939 and, at the end of World War II, it was largely used in those regions where diseases transmitted by insects, such as malaria, typhus fever and yellow fever, were widespread. Moreover, once its usefulness was discovered in the fight against insects harmful for crops, it allowed to increase agricultural production in the immediate post-war period with a consequent rapid recovery of world economy. The problems related to DDT are given by its long decay times that maintain unaltered for years its toxic characteristics. By the time when this insecticide was banned and substituted by other substances, 25% had been carried by rainfall and rivers into the oceans while a great quantity was circulating in food chains.

Pollution Sources

Agricultural Pollution

Agricultural pollution originates from the introduction of chemical fertilizers (rich in phosphates and nitrates), pesticides (insecticides and weed killers) and manure from stables in river flows and in the soil. The discharge of chemical fertilizers in rivers, lakes and seas enhances the eutrophication phenomenon. The introduction of pesticides poses the most serious threat as these are not very biodegradable, they deposit and concentrate in river flows destroying all forms of life.

A greater attention from agricultural operators could substantially reduce this form of pollution that is particularly dangerous as it can regard also aquifers. To prevent this, the contribution of nitrates must be reduced and must be favored the use of natural manure as also must be reduced the use of pesticides introducing biological fight to avoid excessive irrigation that leaches soil and makes necessary the use of fertilizers.

Industrial Pollution

Pollution of industrial origin is caused by the discharge of toxic and non biodegradable substances coming from industrial processing such as cyanides originating from industries producing pesticides and weed killers, cadmium originating from companies producing batteries and accumulators, and chromium as a leftover of plating and tanning industries.

Industrial pollution can derive from the discharge of water used in productive processes that contains high amounts of solid dissolved substances coming from leaching of solid waste landfills carried out by rainwater or by accidental breaks of tanks and/or pipes transporting very polluting products that flow directly into rivers or disperse in the soil and subsoil and eventually reach aquifers. To reduce industrial pollution it's necessary to purify water through filters and treatment tanks before discharging it and favoring, when possible, natural substances in purification processes.

Thermal Pollution

There is also another form of industrial pollution of water that doesn't regard the content of polluting substances but temperature: thermal pollution. Industries, in fact, pour into the sea or into rivers hot water used for their processings.

Cooling water is withdrawn from seas, lakes and rivers at a certain temperature and after use is returned at a higher temperature. The temperature rise in water bodies causes the alteration of aquatic ecosystems and the variation of vital processes. Moreover, it can lead to the death of bacterial fauna which is useful in self-purification processes of water and, in most serious cases, it can also lead to the death of a great number of fish. To reduce the negative effects caused by the discharge of cooling water, hot water produced by domestic heating or for the breeding of species requiring high temperatures should be reused.

Domestic Pollution

Domestic water pollution is produced by the discharge of domestic sludge containing organic substances and soaps.

These substances generally pour into superficial river flows but sometimes reach aquifers. It's possible to reduce wastewater pollution thanks to purification. Discharges are channeled from sewers to treatment stations to abate pollutants before discharging water into rivers and seas. These purification systems, though, aren't always into force and, moreover, even where discharges are gathered and channeled can occur breaks or inefficies of septic pits, pipes or treatment plants that cause leakage of polluted water. Also phytopurification, which uses specific plants that work as biological filters capable of reducing polluting substances, can be employed to abate pollutants present in wastewater.

Hydrocarbon Pollution

It's caused mainly by accidents on oil platforms and ships used for hydrocarbon transport but also by discharging into the sea of water used to wash tanks of tanker vessels. Crude oil and petroleum products form a waterproof film on water that prevents the exchange of oxygen between atmosphere and water causing damages to marine flora and fauna.

Nowadays during transport over sea are used "double-hull" tankers to avoid leaks in case of accidents. The best international practices are adopted with regards to oil plat-forms to face or eventually adequately deal with any type of inconvenience.

Sea and the Ocean Pollution

The use of seawater and the exploitation of marine resources may cause serious dam-ages unless they are carried out in a way that guarantees its sustainable use, i.e. that guarantees compatibility between the marine ecosystems and man's activities. Unfor-tunately, in many cases, since the antiquity the sea has been wrongly considered as a huge dumping ground in which all waste and dirt could be freely thrown. And it is still considered as such by poorly educated summer tourists, especially those who use sail-ing or motorboats and throw all their waste into it.

Today, the main causes of pollution in the seas and oceans can be:

- Pollutants from human activities discharged into the rivers and carried to the sea (degradable and non-degradable organic materials from urban waste, organic products of agricultural origin, such as plant chemicals and fertilizers, pollutants from industrial waste) oil spilled by oil tankers, fol-lowing accidents or improper practices when cleaning tanks or discharging ballast water;

- Radioactive substances: released during nuclear tests, by now stopped in all countries, and when producing atomic fuels;

- Overheating of coastal waters, due to hot water coming from industrial cooling plants;

- Excessive exploitation of fishing resources (too much fishing), that causes the fishing populations to decrease or even disappear;

- Uncontrolled urban development along the coasts and uncontrolled and mas-sive seaside tourism;

- Discharge of nuclear and toxic waste;

- Discharge of plastic containers and other non-biodegradable solid waste.

Heavy Metal Pollution

The most dangerous are: cadmium, lead, mercury that can be harmful to human health even in very low concentrations, as well as being highly toxic and non-degradable. They accumulate in those organisms that occupy the highest levels in the food pyramid: mer-cury pollution in the sea provokes the concentration of this metal in fish and the organ-isms that eat the fish, including men. Mercury that derives from the dumping of indus-trial waste and that reaches the sea is ever lasting and continues its cycle by passing from one organism to another through the food chain.

Euthrophication of the Sea

Oxygen, light and mineral substances are very important for the sea, as they allow organisms to develop. These nutrients melt into the water and their excessive presence makes the sea particularly rich in organisms. In fact the result is an intense growth of algae and aquatic plants that develop rapidly, altering the balance of the ecosystem. Herbivores that eat algae and plants are not enough and do not manage to control these vegetal populations, that forms a large quantity of decomposing material as they die. The decomposition and fermentation of dead organisms means consumption of oxygen, which is less and less available in the environment for those organisms that need it to survive. As a consequence the number of organisms drastically drops.

This situation might occur in the Adriatic Sea, where the Po river waters collect the agricultural, industrial and urban waste coming from the plain of the Po. These polluted waters are rich in nutrients and in the summer of 1989 a phenomenon of euthrophication occurred in the Adriatic sea, which was largely covered by a layer of mucilage produced by unusually growing algae.

Preservation of the Water Resource

Avoid Polluting

In many countries to avoid pollution of water of industrial, domestic and agricultural origin in recent years have been introduced more and more restrictive laws that commit companies and public administrations to give particular attention to prevention, control and reduction of water pollution. New technologies and new products have hence been studied and introduced to allow to produce goods and services limiting and eliminating completely water pollution. Also many international organizations, including the European Commission, have dictated a set of simple recommendations for a sustainable management of water resources. Recommendations range from undertaking reforms of institutions that govern water resources to the definition of an adequate price for water to promote a more cautious and less waste oriented use for water. Sustainable use of water, infact, is based also on waste reduction or its recycling in productive cycles: these practices can increase availability and improve quality of water existing in a given area.

The problem remains, instead, in those countries where these laws haven't been adopted yet or where serious controls are not undertaken on their application. In this case it's desirable that, being aware about water pollution that often is a supranational problem and not only a local one (if pollution enters in the water cycle it can spread also to substantial lengths), industrialized countries should find efficient ways to transfer clean technologies and adequate environmental laws also to poor countries that don't apply them either due to cost issues or because necessary knowledge and training are lacking. This kind of behavior would lead us closer to a situation of sustainable use of this natural resource at a world level.

Need for Water Purification

Once it's used, water is returned strongly deteriorated. It contains, infact, many polluting substances (for example leftovers of detergents used to wash dishes or clothes) or other organic substances (for example human excrements). In many countries (unfortunately not all yet) this water is gathered from the sewage system and sent to a purifier that eliminates or reduces at levels compatible with the health of the environment, concentration of polluting substances; water is finally poured again into natural water flows (rivers and lakes) to return into the sea.

Phytodepuration

In the last decades is being undertaken a "biotechnological solution" which is capable of removing polluting agents from water: phytodepuration is based on self-purification capacity of aquatic ecosystems through physical, chemical and biological processes carried out by vegetal organisms and bacteria. Plants involved are macro and microphytes which are specifically selected according to some characteristics as their capacity to adapt to the environment which needs to be decontaminated and their rapid growth with formation of biomass; in any case, the species employed for phytodepuration are water plants or hygrophilous plants which grow in moist environments. In particular, according to the type of phytodepuration system, which is under construction, are used different types of floating, submerged and emergent microphytes alone or in group.

The type of phytodepuration system established depends on the direction of the water flow. Surface flow systems consist of tanks or channels from 40 to 60 cm deep and recreate an ecosystem similar to ponds covered by floating hygrophilous plants. In sub-surface flow systems, instead, flowing water isn't in contact with the atmosphere and an inert stand is inserted in the tanks where the roots of macrophytes will grow. Water flows under the inert stand to favour movement in the tank, which is 70-80 cm deep and is inclined.

Phytodepuration systems are a valid alternative to wastewater treatment for small rural communities and seasonal sewers as those of camping sites, hotels and holiday villages or for the treatment of industrial wastewater, percolates coming from landfills and run-off water coming from roads and motorways. Construction costs are very variable but, anyhow, are never higher than those of conventional depuration plants whereas management costs are incredibly low as energy consumption can even be nonexistent.

Sustainable Management of Water

Possible actions to manage responsibly such an important resource as water could be many, here are some examples:

- Purification of polluted water: It's possible to restore a lake, as occurred in Switzerland. If the lake is acidified, for example, emission of carbonates neutralizes acidity of polluting substances.

- Pollution prevention: Prevention is indispensable for those water tanks that are not recoverable as aquifers or oceans. In Italy exist different laws referred to pollution prevention including: ban on the use of atrazine herbicides, rationalization of herbicides, fertilizers and pesticides, analysis of potable water, ban of discharging dangerous substances directly or indirectly in aquifers.

- Prevention of the waste of individual water: Paying little attentions in everyday life, we can all commit to saving of this precious resource.

Effluent

Water scarcity is the major problem that is faced all across the world. Although 2/3rd of the earths crust is made up of water but all this water is not available for drinking and for other human activities as either it is locked in the form of ice or present in the form of vast saline oceans and seas. It has been found out that 97% of the total water is salty that is of no use to human and animals (except marine animal) and the remaining three percent is available as freshwater. More than half of this three percent is locked in glacier and less than 0.01% is available as fresh water. So water resources are less as compare to human demand for water.

Above this, the major part of water that can be consumed is getting polluted because of human activities. This polluted and untreated water is causing abundant water borne diseases. Then the world is facing a huge climatic change, which is further aggravating the water problem. Some of the regions are getting more rainwater than earlier and some are getting almost negligible. Experts even believe that the next World War would not be for oil or land but it will be for water.

Also because of improper use of water and lack of water treatment, the problem of water crisis will further increase where 884 million people are already not getting easy access to safe drinking water. And a further 2.5 billion people are getting difficult access to water for disposable and sanitation. Agriculture is also overusing and polluting the ground water thus depleting the natural source of water. So here water treatment plants will play important role.

Water crisis at present is the biggest problem according to the United Nations. Almost 25 countries of Africa, parts of China, Peru and Brazil in Latin America, some parts of Middle East like Iran, Chile, Mexico, and Paraguay are some of the countries that are facing the water crisis. Even other parts of the world are facing the varied levels of the water crisis. Because of acute shortage of water, the food problems are getting aggravated. About 40 million people in Africa are facing the problem of food shortage. It is expected that if the similar conditions will persist then there will be 500 million till 2025 who will suffer from these problems. Nature has its role but the major water problem is arising because of its increasing consumption and faulty usage. Major chunk of the problem can be solved if the wastewater treatment is taken very seriously and precautions at every step are taken to improve the water quality.

The Supply and Demand

The demand for fresh clean water delivered to our homes is ever increasing as more and more residential homes are being established. Although 70% of the world is covered with water, only 1% is fresh water and thus raises a need to recycle wastewater to satisfy our needs. Efforts to continuously recycle wastewater are always stressed upon, as a shortage would mean a disaster in heavily populated areas. Governments have committed billions towards research and development to such projects. Fresh water is also needed in agriculture. The demand for water in this sector is very high, as farmer needs fresh water for crops and cattle. Therefore, stresses the demand for sewage water treatment plants to be built. Water from rivers and lake are inadequate to provide water for farm and residences alike.

The supply of sewage water treatment facilities is slowly dwindling. A crisis may arise from a lack of sewage water treatment plants, as this would greatly reduce the supply of fresh water. Governments are desperately trying to keep up with the pace of development of the population but are slowly lagging behind. The supply of fresh water will be adequate for the next few years but if the development of sewage water treatment plant continues at its current pace it without a doubt a shortage would take place. Furthermore, the agriculture industry will be greatly impacted and the supply of food will go down as well leading to famine due to a shortage of water.

The Prospects of Wastewater Industry

The wastewater treatment industry most probably will be successful in future due to presence of new wastewater treatment technologies. Advanced Immobilized Cell Reactor technology is one of the new technologies, which immobilize the organisms such as bacteria in the pores of the carbon matrix. This process can avoid the immobilized organisms from shock load application as the diffusion of the pollutants from bulk fluid phase to organisms follows Fick's law.

Through conventional biological wastewater treatment, infinite electrical energy and vast land area are being consumed. Besides that, a huge investment in electromechanical equipment is involved which will bring about a huge total cost of operation. Generally, the total cost of operation for new technology is lowered compared to the convention technology. By using the new technology, the total cost of operation can be cut down to approximately 50 percent of the total cost of conventional treatment.

Furthermore, the biological oxygen demand and chemical oxygen demand are reduced by 94 percent and 90 percent respectively. Oxygen consumption in the new technology is lower than in conventional technology. The oxygen gas is supplied in the form if compressed air from the bottom of the reactor. Both liquid and gas streams are in counter-current direction, which facilitates the oxidation of dissolved organics and desorbs the converted products. This is to make sure the activated carbon keeps up its activity throughout the process.

Moreover, with all those new wastewater technologies such as Advanced Oxidation Process, NERV (Natural Endogenous Respiration Vessel), Wet oxidation and many others processes, wastewater treatment can be done efficiently. For instant, through the new technologies less land is required to use to build plant; the power consumption is lowered. Besides that, the requirement for electrical and mechanical equipment is lower compared to conventional technology.

In a conclusion, wastewater treatment industries have a good prospect in the future with the help of new technologies. By using all those new technologies, wastewater treatment can be done efficiently with lower overall lifecycle costs, lesser energy and equipment needed. We are sure that there is more new technologies will be invented in order to improve the wastewater treatment.

The Impact on the Environment

When the wastewater is mixed with the waste materials such like garbage, household waste, toilets liquid and disposable things, the resulting product called sewage or wastewater. This sewage water is normally will undergo a few process before it is release to the environment but there are still some impact on the environment. One of the impacts on the environment is agricultural impact. The sewage water contains salts which is soluble that may accumulate in the root zone with possible harmful effect on soil health and crop yield. The physical and mechanical properties of the soil, such as dispersion of particle, stability of aggregates, soil structure and permeability are very sensitive to the types of exchangeable ions present in irrigation water. Thus, when effluent use is being planned, several factors related to the soil properties must be taken into consideration. On the other hand the effect of dissolved solids in the irrigation water on the growth of plants is also another aspect of agriculture, which we have to concern. Dissolved salts increase the osmotic potential of soil water and increase the osmotic pressure of the soil solution, which increases the growth, and the yield of most plants decline progressively as osmotic pressure increases. In addition the one of the environment impact is ecological impact where the drainage water from wastewater irrigation schemes drains particularly into small confined lakes and water bodies and surface water, and if phosphatesin the ortho phosphate form are present, the remains of nutrients may cause eutrophication. Here the overloading organic materials resulting in decrease in dissolved oxygen may lead to changes in the composition of a aquatic life such as fish deaths and reduced fishery. The eutrophication potential of wastewater irrigation can be assessed using biological indices, which in turn can be qualified in monitory units using economic valuation techniques. The hidden impact on the environment is the increase on the production of green house emissions. The large agriculture reuse project might cause to the environmental externalities associated with pumping water uphill, which emits greenhouse gas. Another impact is on the health. The sewage water contains pathogenic microorganisms like bacteria, viruses, protozoan and parasitic worms, the diseases and signs related with such infection are also diverse including typhoid,dysentry and cholera, diarrhea and vomiting. the concentration of he pathogens in waste water is dependent on the source population and the susceptibility to infection varies from one population to another. So basically he waste water is actually harm for the nature even though its treated and release to the environment so as a human being we should not dispose the waste into the water thus our water will be clean and the cost of the treatment can be reduced.

The Processes Involved in Wastewater Industry

Pre-treatment consists of three sub-stages which are Screening, Grit Removal and Fat and Grease Removal. Pre-treatment is done to remove materials which are easily collected such as debris, leaves and trash which would damage or clog up pumps and skimmers of the primary treatment.

Screening is used to remove large objects such as leaves, twigs and cans in the sewage stream. This is normally done with a giant mechanical rake bar, which is automated. The rake bar revolves around a central axis at a rate varying on the accumulation and flow rate of the sewage stream. The screens vary in sizes to optimize solid removal. Objects accumulated are collected and disposed in landfills.

Grit is a minute granule such as sand or stone. The wastewater is channeled to a chamber where to velocity of the water is adjusted so that the grit would settle at the bottom of the chamber. Grit may cause damage to the pumps or other equipment. Grit removal may not necessary in smaller plant.

Fat and grease are groups of compounds, which are generally insoluble in water. The fat and grease are normally found floating on the surface of the water. In some plants, the fat and grease are removed by using skimmers to collect the fat and grease on the surface of the water in a small tank. However this can also be done in the Primary treatment stage in the same manner.

Primary Treatment

Primary wastewater treatment is the second step in the wastewater treatment process ahead of the preliminary treatment of a headwork's, involves the physical separation of suspended solids from the wastewater flow using primary clarifiers. The objective of primary treatment is the removal of settles able organic and inorganic solids by sedimentation, and the removal of materials that will float (scum) by skimming. Approximately 25 to 50% of the incoming biochemical oxygen demand (BOD_5), 50 to 70% of the total suspended solids (SS), and 65% of the oil and grease are removed during primary treatment. Some organic nitrogen, organic phosphorus, and heavy metals associated with solids are also removed during primary sedimentation but colloidal and dissolved constituents are not affected. The effluent from primary sedimentation units is referred to as primary effluent.

On the other hand, primary treatment is the minimum level of reapplication treatment required for wastewater irrigation. It may be considered sufficient treatment if the wastewater is used to irrigate crops that are not consumed by humans or to irrigate orchards, vineyards, and some processed food crops. However, to prevent potential nuisance conditions in storage or flow-equalizing reservoirs, some form of secondary treatment is normally required in these countries, even in the case of non-food crop irrigation. It may be possible to use at least a portion of primary effluent for irrigation if off-line storage is provided.

Primary sedimentation tanks or clarifiers may be round or rectangular basins, typically 3 to 5 m deep, with hydraulic retention time between 2 and 3 hours. Settled solids (primary sludge) are normally removed from the bottom of tanks by sludge rakes that scrape the sludge to a central well from which it is pumped to sludge processing units. Scum is swept across the tank surface by water jets or mechanical means from which it is also pumped to sludge processing units.

Secondly Treatment

The secondary treatment in this sewage treatment is one of the most important part in this process. This process is basically designed to remove the waste product from the sewage. This system is also classified as fixed-film or suspended-growth systems. The secondary treatment contain a few processes, the 1st process is activated sludge. This activated sludge is majority from the plants, which encompass the variety of mechanisms and processes that use dissolves oxygen to promote the growth of biological flock that substantially removes organic material. This process basically changes the ammonia to nitrite and nitrate and ultimately to nitrogen gas. The 2nd process is this treatment is the Surface-aerated basins also known as Lagoons. This process basically removes the BOD from the sewage water. In an aerated basin system, the aerators provide two functions: they transfer air into the basins required by the biological oxidation reactions, and they provide the mixing required for dispersing the air and for contacting the reactants (that is, oxygen, wastewater and microbes). However, they do not provide as good mixing as is normally achieved in activated sludge systems and therefore aerated basins do not achieve the same performance level as activated sludge units. The biological oxidation in the Surface-aerated basins is sensitive to the temperature and the rate of reaction increase with the temperature. The suitable temperature for this process is in between 0° C and 40° C. Besides that the constructed wetland is one of the process also. This process is a process, which cleans the drainage of animals and used to recycle the wastewater. The constructed wetland are known to be highly productive systems as they copy natural wetlands, called the "Kidneys of the earth" for their fundamental recycling capacity of the hydrological cycle in the biosphere and they provide a high degree of biological improvement but depending on design. The next process is the filter beds which is knows as oxidizing beds are used where the settled sewage liquor is spread onto the surface of a bed made up of coke, then liquor is typically distributed through perforated spray arms, then distributed liquor trickles through the bed and is collected in drains at the base, and the biological films of bacteria, protozoa and fungi to reduce the organic content. The next process is the Biological aerated filters are a combine filtration with biological carbon reduction, nitrification or denitrification. It's a dual processer in purpose of to support highly active biomass that is attached to it and to filter suspended solids. Carbon reduction and ammonia conversion occurs in aerobic mode and sometime achieved in a single reactor while nitrate conversion occurs in anoxic mode. This process is operated either in up flow or down flow configuration depending on design specified by manufacturer. In addition the Rotating biological contactors are the next process in this secondary treatment. This is actually a secondary mechanical treatment system, which is capable of withstanding surges in organic load.

The rotating disks support the growth of bacteria and micro-organisms present in the sewage, which break down and stabilize organic pollutants. Oxygen is obtained from the atmosphere as the disks rotate. As the micro-organisms grow, they build up on the media until they are sloughed off due to shear forces provided by the rotating discs in the sewage. Effluent from the system is then passed through final clarifiers where the micro-organisms in suspension settle as sludge. The sludge is withdrawn from the clarifier for further treatment. After that the membrane bioreactor combine activated sludge treatment with a membrane liquid-solid separation process. The component on this system uses low pressure for microfiltration or ultra-filtration membranes and eliminates the need for clarification and tertiary filtration. The elevated biomass concentration in the system process allows for very effective removal of both soluble and particulate biodegradable materials at higher loading rates. The final process in this secondary treatment is the secondary sedimentation where the process is to settle out the biological flock or filter material through a secondary clarifier and to produce sewage water containing low levels of organic material and suspended matter.

Tertiary Treatment

The main purpose of the tertiary treatment is to ensure that the treated water, which is to be released on to the environment, is biologically accepted by all other fresh water organisms such as weeds and algae. This part of the treatment includes processes like physical water treatment, lagooning, and excessive nutrient removal processes.to ensure that the discharged water is raised in effluent quality before proceeding to the final stages.

In physical water treatment, much of the residual suspended matters are removed using only physical processes such as sedimentation method and the infamous filtration method. In the sedimentation method, the water is place in a certain tank to allow all the remaining heaver objects to sink down to the bottom of the container. After few hours went most of the dense object are separated from the water, the cleared effluent or waste stream is removed. Sedimentation is one of the most common methods, quite often used at the beginning and the end of many water-treating processes. Another physical method that is commonly used in the sewage water treatment system is the filtration method. In filtration, the water is allowed to pass through filters to separate the contaminating solids from the water. Sand filter is a common filter used in this process. In a number of wastewater treatment methods, semi-solid contaminants like grease and oil are allowed to float on the surface of the water, and then they are physically removed.

Besides the in lagooning where lagoon is a stationary system having a continuous flow: several ponds working in parallel in which the inlet flow and the outlet flow are equals form lagoon plants. The lagooning technique is a natural and very efficient technique that consists in the accumulation of wastewater in ponds or basins, known as biological or stabilization ponds, where a series of biological, biochemical and physical processes take place. In these ponds or lagoons, certain types of the microorganism are actually supported as these "biological agents" help in treating the water further by removing the fine particulates. These types of biological ponds are usually classified as anaerobic ponds or oxidation ponds depending on the shape, depth, and organic rate, level of treatment of that particular lagoon itself.

The excessive nutrient removal is the most viral step in the last stages of the water treatment before the water is released to the environment. When the previously treated water comes to this area of the system, the nutrients level mainly nitrogen and phosphorus in the water is checked. Where when found in excess, the excessive nutrient removal step is carried out. This is because if the unchecked water supply is to be released into the natural water system (river, pond, etc.) it will cause a sudden increase in the native microorganism population of that certain water system. Some of the native microorganisms, which are commonly found in ponds today, are usually weeds, algae, and cyanobacteria. Therefore, after the sudden rapid grown in the population of these microorganisms, the number of algae for example, becomes unsustainable which causes most of them to die and eventually decay. The following decay process would substantially increase the biochemical oxygen demand (BOD) in that particular pond, which would cause the other fresh water living organism such as fishes to die as well.

Firstly, the nitrogen removal process is carried out. There are various methods of removing nitrogen, each with advantages and disadvantages. However, the biological treatment method is used most commonly. With this method, organic nitrogen and ammonia nitrogen is converted into nitrous and nitrate nitrogen in an aerobic environment, and is dispersed into the atmosphere as anaerobic nitrogen gas. Therefore the gas is removed from the water and released to the atmosphere. And as there is no secondary pollution, this can be called an effective method.

In the removal of phosphorus is usually carried out using a method called enhanced biological phosphorus removal (EBPR). The first process in EBPR is the mainstream biological treatment process. Where the utilizing of aerobic and solids separation zones and the provision of return activated sludge is carried out. The next is a first side stream process for anoxic/anaerobic "selection" of desirable BPR organisms such as the polyphosphate accumulating organisms. Finally, a second side stream process serves to ferment organic material in some of the return activated sludge to produce food utilized in the first side stream selection process. The system permits the three processes to be separated from each other by creating two side streams, allowing all three processes to be controlled separately and optimized in satisfying their own specific goals. Besides this biological method, the removal of phosphorus can also be done via chemical precipitation, usually with salts of iron, aluminum, or lime. Chemical precipitation is usually more reliable, easier to operate, and requires smaller equipment footprint than biological removal. But the main back draws of this chemical method are that it may form excessive sludge production as hydroxides precipitates and the chemical used in this method might be considered expensive.

Disinfection

The main purpose of disinfection in the wastewater treatment is to provide a degree of protection from contact with infectants and pathogen organisms, which will cause waterborne diseases such as cholera, dysentery and hepatitis. Disinfection is also used to reduce the load of microorganisms in the wastewater to be discharged to the environment. Primary, secondary and even tertiary treatments do not fully remove the incoming waste load and microorganisms in the water stream and as a result, many microorganisms still remain in the wastewater. Therefore, various methods of disinfection are introduced such as chemical methods, physical methods and biological methods.

The effectiveness of disinfection depends on different factors including the quality of wastewater being treated, disinfectant dosage, type of disinfection being used and others. For instant, cloudy wastewater will not be treated efficiently due to less contact time between ultraviolet light and microorganisms. These microorganisms are shielding by those solid matters in wastewater stream and it reduces the contact time. Generally, long contact times, high concentration of disinfectant and optimum temperature and pH value will increase the effectiveness of disinfection.

Chlorination is one of the chemical methods, which is commonly used for disinfection in the wastewater treatment. It is widely used through the world due to its low cost and long-term history of effectiveness. Chlorine can be applied in two general ways, liquid and gas. Chlorine in gaseous form is generally added to the wastewater stream rather than liquid form, which is also known as hypochlorite because the former costs lesser than the latter. When chlorine dissolves in pure water, hypochlorous acid is formed followed by hypochlorites, which are known as "free" residual chlorines

Chlorine is an extremely active oxidizing agent, which will react with many other substances in the water stream. For instant, it reacts rapidly with such compounds as hydrogen sulfide, ferrous iron and manganese, which found in industrial wastewater. However, if all of the chlorine is consumed in these reactions, no disinfection will result. Hence, to accomplish disinfection, sufficient chlorine is added into wastewater stream to satisfy the chlorine demand and produce residual chlorine, which will destroy bacteria.

There are few factors, which will affect the effectiveness of chlorination. Among the factors are pH, temperature, turbidity, control system and many others. However, chlorination brings some disadvantage to environment. Chlorination of residual organic material can generate chlorinated-organic compound, which may be harmful to the environment. Those residual chlorines are toxic to aquatic species; therefore, dechlorination is needed, adding to complexity and cost of treatment.

However, chlorination becomes less favored as disinfectant due to rising cost and it had found to be toxic to aquatic species. As a result, ozone and ultraviolet begin to be used as disinfectant. Ultraviolet (UV) light is more environmental friendly to be used as no chemicals are used and leave no toxic residual. Ultraviolet radiation and damages the genetic material of microorganisms, destroying their ability to reproduce. Before pass through the UV disinfection unit, the wastewater must pass through an advanced pretreatment component. Wastewater flows in the stream parallel to the UV light in a thin film in order to increase the contact time.

To increase the effectiveness of the UV light, the UV radiation must come in direct with pathogen organisms and other microbial in the wastewater stream. The effectiveness of a UV disinfection system is affected by few factors including characteristics of the wastewater, the contact time, intensity of UV radiation and many others. Turbidity, flow rate of water stream and suspended solids are also playing an important role in UV disinfection. These factors must be kept at low levels to ensure proper treatment.

Disinfection of wastewater, primarily by chlorination, has played an important role in the reduction of waterborne disease. However, there are more new disinfection processes are being developed in order to maximize the effectiveness of disinfection.

Acidification of Water

Acidification of freshwaters was a problem that was first identified in Scandinavia during the early 1970s, at which time many scientific studies were initiated. Since then the concerns that were voiced have been justified, and now thousands of lakes and rivers are known to be acidified. Areas that are most susceptible to acidification have an unreactive geology such as granite and a base-poor soil. Areas that are affected by acidification include Scandinavia, Central Europe, Scotland, Canada, and the United States. Lakes and streams that are generally regarded as acidified are very nutrient poor waters draining unreactive geology.

Ample evidence from chemical and biological studies of typical lakes proves that increased acidification has taken place. Diatom shells from lake sediments have allowed the course of acidification to be charted back through time. Diatoms are microscopic algae, which live free floating in the water or attached to surfaces. They have hard shells of silica, which are characteristic of each species. Diatoms are very sensitive to acidity, and their occurrence and proportions give good indications of pH levels. Evidence suggests that rapid acidification has been taking place at some sites for at least 100 years and is still occurring today.

Process

Acid rain can enter the watercourse either directly or more usually through the catchment. If the catchment has a thin, base-poor soil then acid water is passed to the lake. If the catchment has alkaline-rich soil then the acid rain is neutralized and so water entering the lake is of low acidity. In areas where a continual supply of base (alkali) cations is not assured then the gradual depletion of the bicarbonate in the lake means that the once stable pH will drop rapidly resulting in an acidified lake. Acidification can also occur in surges after snowmelt or drought; the first 30% of snowmelt can contain 50 - 80% of the total acids in the snow. During drought conditions sulphur dioxide (SO_2) deposition onto the soil is reduced to sulphur and hydrogen; this is then re-oxidized in combination with rainwater to form acids. This is termed an acid pulse.

Effects

The effects of freshwater acidification are as follows:

- Carbon source changes from carbonate (HCO_3) to carbon dioxide (CO_2).
- Release of toxic metals.
- Phosphorous is retained.
- Freshwater fauna and flora gradually changes.
- Short-term pH depressions have direct toxic effects on susceptible organisms.

The onset of acidification brings about a clearer bluer water body due to the precipitation of humic substances. Whilst total biomass remains largely unchanged, the diversity drops considerably. Many algal species disappear, but some green filamentous algae are capable of mass proliferation in the extreme environment. The number of macrophytes in and around the water decreases, with Rush becoming the dominant species. White Sphagnum moss may invade lakes and form a thick green carpet over the bottom of the lake on account of the clearer waters allowing more light to reach the moss. In some acidified lakes the abundance of larger plants (called macrophytes) has decreased, sometimes accompanied by increased abundance of a moss known as Sphagnum. In itself, proliferation of Sphagnum can cause acidification, because these plants efficiently remove cations from the water in exchange for H^+, and their mats interfere with acid neutralizing processes in the sediment.

Causes

Acidification takes place most readily in areas where the natural geology is slightly acidic. Upland regions that have been subject to land-use changes over the last few decades are showing the signs of acidification.

Several factors affect acidity:

Natural:

1. Action of atmospheric carbonic acids.

2. Formation of organic acids by humus podsolization.

3. Podsolization.

Land-use changes:

4. Livestock introduction into the catchment.

5. Use of nitrogen fertilizer.

6. Increased efficiency of drainage.

7. Dry deposition of air pollutants.

8. Wet deposition of sulphuric and nitric acids.

It will be a combination of the above factors that will lead to freshwater acidification. Natural acidification has been taking place since the last ice age, although the recent rapid acidification of many of the lakes cannot be attributed to natural causes.

Surface water Acidification

There is very little direct chemical evidence that shows that surface waters have acidified, both in the UK and worldwide. The problem of acid rain was only identified once emissions of acidifying pollutants were on the decline, and monitoring data largely fails to capture the trend of increasing acidity. However, evidence for the acidification of surface waters has been obtained from pH reconstructions based on diatoms in lake sediment cores. This palaeolimnological data has provided the key evidence linking historical emissions of acidifying pollutants with the acidification of surface waters.

In the early 1980s the link between acid deposition and the acidification of surface waters was not universally accepted. Alternative hypotheses for the cause of the acidification were proposed:

- Natural acidification processes

- Changes in catchment management

- Afforestation of upland soils

A major project of palaeolimnological work was funded by the Surface Water Acidification Programme (SWAP) and later by the then Department of the Environment to evaluate these alternative hypotheses. Sites in Norway, Sweden and the UK were chosen to answer specific questions about the timing, extent and cause of surface water acidification. A range of techniques were employed to answer these specific questions:

- Analysis of the remains of diatoms and other biological groups

- Trace metal analysis

- Fly-ash particle analysis

The overall conclusion was that acid deposition was the main cause of the recent acidification of surface waters, although natural acidification had increased the sensitivity of many sites to the effects of acid deposition. Afforestation had also led to stronger acidification at those site in areas receiving high levels of acid deposition.

Restoration of Acid Waters

The only sure way to prevent further acidification of other susceptible water bodies is to reduce the emissions of acid pollutants. There is a relationship between sulphur emissions, deposition, sulphur in run-off and loss of alkalinity. If acidification of soils and freshwaters is to be prevented then sulphur deposition rates need to be reduced further. The technical means are available to reduce emissions, such as flue gas desulphurisation, low NOx burners, use of low sulphur coal and oil and increasing energy efficiency. At present the main way of reversing acidification in freshwaters is liming the water body or its surrounding catchment. The main liming method is to add the lime directly to the water body. However in the cases of certain lakes where the turnaround is very quick, the lime is added to the catchment. This has disadvantages though; the main one being that the lime can have an adverse effect on wetland species of plants.

The advantages, however, are that the effects are longer lasting and metals are prevented from leaching into the lake water from the soil. The effects of liming are almost entirely favorable within the lake. The alkalinity of the limed lake is increased, the pH increased and heavy metal concentrations decrease back to within safe limits for fish life. The number of species of fish, benthic animals and plankton increases, as does biomass production.

Ocean Pollution

Oceans are the largest water bodies on the planet Earth. Over the last few decades, surplus human activities have severely affected the marine life on the Earth's oceans. Ocean pollution, also known as marine pollution, is the spreading of harmful substances such as oil, plastic, industrial and agricultural waste and chemical particles into the ocean. Since oceans provide home to wide variety of marine animals and plants, it is responsibility of every citizen to play his or her part in making these oceans clean so that marine species can thrive for long period of time.

Mining for materials such as copper and gold is a major source of contamination in the ocean. For example, copper is a major source of pollutant in the ocean and can interfere with the life cycles of numerous marine organisms and life.

Causes of Ocean Pollution

There are various ways by which pollution enters the ocean. Some of them are:

Sewage

Pollution can enter the ocean directly. Sewage or polluting substances flow through sewage, rivers, or drainages directly into the ocean. This is often how minerals and substances from mining camps find their way into the ocean.

The release of other chemical nutrients into the ocean's ecosystem leads to reductions in oxygen levels, the decay of plant life, a severe decline in the quality of the sea water itself. As a result, all levels of oceanic life, plants and animals, are highly affected.

Toxic Chemicals From Industries

Industrial and agricultural wastes are another most common form of wastes that are directly discharged into the oceans, resulting in ocean pollution. The dumping of toxic liquids in the ocean directly affects the marine life as they are considered hazardous and secondly, they raise the temperature of the ocean, known as thermal pollution, as the temperature of these liquids is quite high. Animals and plants that cannot survive at higher temperatures eventually perish.

Land Runoff

Land runoff is another source of pollution in the ocean. This occurs when water infiltrates the soil to its maximum extent and the excess water from rain, flooding or melting flows over the land and into the ocean. Often times, this water picks up man-made, harmful contaminants that pollute the ocean, including fertilizers, petroleum, pesticides and other forms of soil contaminants. Fertilizers and waste from land animals and humans can be a huge detriment to the ocean by creating dead zones.

Large Scale Oil Spills

Ship pollution is a huge source of ocean pollution, the most devastating effect of which is oil spills. Crude oil lasts for years in the sea and is extremely toxic to marine life, often suffocating marine animals to death once it entraps them. Crude oil is also extremely difficult to clean up, unfortunately meaning that when it is split; it is usually there to stay.

In addition, many ships lose thousands of crates each year due to storms, emergencies, and accidents. This causes noise pollution (excessive, unexpected noise that interrupts the balance of life, most often caused by modes of transportation), excessive algae, and ballast water. Often times, other species can also invade an ecosystem and do harm to it by interrupting the life cycles of other organisms, causing a clash of nature that has already been damaged by the overflow of pollution.

Ocean Mining

Ocean mining in the deep sea is yet another source of ocean pollution. Ocean mining sites drilling for silver, gold, copper, cobalt and zinc create sulfide deposits up to three and a half thousand meters down in to the ocean. While we have yet the gathering of scientific evidence to fully explain the harsh environmental impacts of deep sea mining, we do have a general idea that deep sea mining causes damage to the lowest levels of the ocean and increase the toxicity of the region. This permanent damage dealt also causes leaking, corrosion and oil spills that only drastically further hinder the ecosystem of the region.

Littering

Pollution from the atmosphere is, believe it or not, a huge source of ocean pollution. This occurs when objects that are far inland are blown by the wind over long distances and end up in the ocean. These objects can be anything from natural things like dust and sand, to man-made objects such as debris and trash. Most debris, especially plastic debris, cannot decompose and remains suspended in the oceans current for years.

Animals can become snagged on the plastic or mistake it for food, slowly killing them over a long period of time. Animals who are most often the victims of plastic debris include turtles, dolphins, fish, sharks, crabs, sea birds, and crocodiles.

In addition, the temperature of the ocean is highly affected by carbon dioxide and climate changes, which impacts primarily the ecosystems and fish communities that live in the ocean. In particular, the rising levels of CO_2 acidify the ocean in the form of acid rain. Even though the ocean can absorb carbon dioxide that originates from the atmosphere, the carbon dioxide levels are steadily increasing and the ocean's absorbing mechanisms, due to the rising of the ocean's temperatures, are unable to keep up with the pace.

Effects of Ocean Pollution

Effect of Toxic Wastes on Marine Animals

Oil spill is dangerous to marine life in several ways. The oil spilled in the ocean could get on to the gills and feathers of marine animals, which makes it difficult for them to move or fly properly or feed their children. The long-term effect on marine life can include cancer, failure in the reproductive system, behavioral changes, and even death.

Disruption to the Cycle of Coral Reefs

Oil spill floats on the surface of water and prevents sunlight from reaching to marine plants and affects in the process of photosynthesis. Skin irritation, eye irritation, lung and liver problems can impact marine life over long period of time.

Depletes Oxygen Content in Water

Most of the debris in the ocean does not decompose and remain in the ocean for years. It uses oxygen as it degrades. As a result of this, oxygen levels go down. When oxygen levels go down, the chances of survival of marine animals like whales, turtles, sharks, dolphins, penguins for long time also goes down.

Failure in the Reproductive System of Sea Animals

Industrial and agricultural wastes include various poisonous chemicals that are considered hazardous for marine life. Chemicals from pesticides can accumulate in the fatty tissue of animals, leading to failure in their reproductive system.

Effect on Food Chain

Chemicals used in industries and agriculture get washed into the rivers and from there are carried into the oceans. These chemicals do not get dissolved and sink at the bottom of the ocean. Small animals ingest these chemicals and are later eaten by large animals, which then affects the whole food chain.

Affects Human Health

Animals from impacted food chain are then eaten by humans which affects their health as toxins from these contaminated animals gets deposited in the tissues of people and can lead to cancer, birth defects or long term health problems.

Ocean Deoxygenation

Oxygen is necessary to support life on Earth. Oxygen makes up about 21% of the air we breathe and half of this oxygen is produced by phytoplankton in the ocean. All aerobic life requires oxygen to produce energy. In water, oxygen is found in a dissolved form and is much more limiting. For this reason, we measure dissolved oxygen in water not in % but in units of umol/kg. The highest concentrations of dissolved oxygen in the ocean are found within the surface mixed layer at concentrations of > 200 umol/kg. Below the surface mixed layer, oxygen concentrations are lower due to respiration by all of the trillions of organisms and bacteria that live in the water column.

Ocean deoxygenation refers to the loss of oxygen from the oceans due to climate change. Long-term ocean monitoring shows that oxygen concentrations in the ocean have declined during the 20th century, and the new IPCC 5th Assessment Report (AR5 WG1) predicts that they will decrease by 3-6% during the 21st century in response to surface warming. While 3-6% doesn't seem like much, this decrease will be felt acutely in hypoxic and suboxic areas, where oxygen is already limiting. "Hypoxic" areas are defined as regions where oxygen limitation is detrimental to most organisms. This threshold differs across the world, but is usually defined as anything below

60 umol/kg. Hypoxic zones have oxygen concentrations 70-90% lower than the mean surface concentrations. "Suboxic" areas are areas where oxygen is so low (less than 5 umol/kg) that most life cannot be sustained and significant biogeochemical changes occur due to altered water chemistry. Suboxic zones have oxygen concentrations 98% lower than the mean surface concentrations. A recent study found that a 1° C warming throughout the upper ocean will result in the increase of hypoxic areas by 10% and a tripling of the volume of suboxic waters. To put this in context, a highly optimistic emissions scenario of atmospheric CO_2 levels of 550 ppm by 2100 would lead to a 1.2° C warming of the upper ocean. Therefore, these declines in oxygen are changes we should be prepared to see.

Lowering of the Oxygen Levels in the Ocean by humans

Oxygen content in the water is dependent on photosynthesis (produces oxygen), animal respiration (uses oxygen), and physical mixing. Ocean warming is reducing global ocean oxygen content through several key mechanisms including:

Stratification impacts: Anthropogenic warming causes surface waters to become warmer and thereby less dense, leading to a more stratified (layered) water column, which reduces mixing. Other impacts of climate change to the water cycle can also lead to a more stratified water column. These include inputs of freshwater to the ocean from rain, river runoff, or melting ice.

Warming effects: As a physical rule, warmer water holds less oxygen. As the surface waters warm due to climate change, the ocean loses its ability to hold oxygen, leading to an oxygen decline.

Biological effects: Changes to the biological use and production of oxygen can lead to changes in oxygen content in the water. Warmer ocean temperatures increase oxygen demand from organisms. Increased nutrient inputs (either through coastal runoff or through upwelling) also lead to more oxygen depletion at mid-depths (100-1000m).

Circulation changes: Changes in ocean circulation are also implicated with some of the observed declines in dissolved oxygen. Slowing circulation and increased upwelling of oxygen-poor deep-water can lead to reductions in oxygen.

Expansion of low oxygen areas due to Climate Change

Low oxygen areas are already present in several parts of the world and are increasing in number, volume, and intensity. Eastern boundary currents and upwelling regions occur on the eastern side of ocean basins and support some of the richest fisheries in the world. However, underlying these productive surface waters are extremely oxygen-depleted waters at 100-1000m depths. These are known as oxygen minimum zones (OMZs), which are midwater regions of the ocean that are naturally low in oxygen due to the combined processes of high oxygen use and limited oxygen replenishment. Already, OMZs make up ~8% of the total oceanic area. OMZs are now expanding both horizontally and vertically due to climate change, resulting in habitat loss for organisms that are sensitive to low-oxygen concentrations. Since the 1960s, the hypoxic area has increased by 4.5 million km2 at depths of 200-700 m in tropical and subtropical waters. Meanwhile, off of California, waters at 200-300m have lost 20-30% of their oxygen in the last 25 years. OMZ expansion is evident in all tropical ocean basins and throughout the subarctic Pacific, making habitat compression an increasingly global issue.

Fisheries impacts from oxygen minimum zone expansion

The expansion of oxygen minimum zones and habitat compression is predicted to impact oceanic commercial fisheries. Already, some species range shifts have been observed due to changes in oxygen content. For example, along the Japanese continental slope decreases in mid-depth oxygen content over the last 60 years have resulted in Pacific cod shifting their distributions to shallower depths. In the tropical Atlantic, blue marlin and tuna have experienced a 15% reduction in vertical habitat between 1960-2010 due to the expansion of oxygen minimum zones. While organisms that are intolerant of low oxygen are losing habitat, other organisms are gaining habitat. Off the West Coast of the US, the Humboldt squid has greatly expanded its range, and the range expansion coincides strongly with areas of significant oxygen declines.

Affects of oxygen loss on marine ecosystems and humans

Oxygen plays a key role in structuring marine ecosystems and controls the distribution of essentially all marine organisms. For this reason, oxygen loss in the oceans will have significant ecosystem-level consequences. Organisms exhibit a great range in tolerances to low levels of dissolved oxygen. Some are highly tolerant, such as jellyfish and squid, while other groups like fish and crustaceans require higher oxygen levels and are highly vulnerable to oxygen declines. This sensitivity has been seen in many systems. For example, off Southern California mid-water fishes, which are crucial components of the ocean food web, declined 63% between periods of high and low oxygen. The vulnerability of fish and crustaceans is also evident by the large fish and crustacean die-offs that have occurred during low oxygen events, including off Oregon's coast and in the Gulf of Mexico. While mortality is a direct impact of oxygen loss, marine organisms can also be indirectly impacted leading to changes in behavior, reproduction, growth and behavior. Together, direct and indirect species-level impacts can lead to significant ecosystem-level consequences such as decreased resilience, stability, and resistance to other anthropogenic stressors such as fishing and pollution. Coastal waters are experiencing significant reductions in oxygen due to nutrient pollution, but these impacts are also exacerbated by climate change. Dead zones, areas where most organisms cannot live due to oxygen limitation, are now reported for more than 479 systems, and since the 1960s, their numbers have doubled approximately every decade. Currently, hypoxia and anoxia are among the most widespread deleterious anthropogenic influences on estuarine and marine environments. As surface waters warm, these low oxygen coastal areas, and the animals that live within them, experience even more oxygen stress. These changes impact the people that depend on these waters for resources.

Need for Attention to Deoxygenation By Policy-Makers

Marine ecosystems provide us with fundamental ecosystems services and changes to these ecosystems leave coastal economies vulnerable. Approximately 470 to 870 million of the poorest people in the world rely heavily on the ocean for food, jobs, and revenues and live in countries that will be most affected by the stacked impacts of warming, acidification, and deoxygenation. Namibia, the Philippines, India, Chile, Peru, and the US are just several of the coastal nations that are vulnerable to the impacts of deoxygenation. Because deoxygenation may have direct economic impacts on marine resources, it warrants attention by policy-makers.

The potential consequences of ocean oxygen loss are profound, but deoxygenation does not act alone. Together, warming, acidification, and deoxygenation present a triple whammy for marine life. Warming and deoxygenation are important synergistic stressors for marine organisms because oxygen requirements vary with temperature, and temperature thresholds are often limited by oxygen availability. Low oxygen waters are also characterized by low pH, meaning that marine organisms are simultaneously experiencing oxygen and pH stress. Meanwhile, decreases in pH may require organisms to use more energy to maintain acid-base homeostasis, however, organismal metabolism may be limited by low oxygen conditions. Due to their interacting effects, warming, acidification, and deoxygenation need to be considered together in order to understand how marine ecosystems will respond to climate change.

Mitigation Strategies

Since climate change is the driving cause of ocean deoxygenation, reducing carbon dioxide (CO_2) emissions is the only real solution. However, certain other actions can help to ameliorate the problem especially at a local level. Lessening other anthropogenic stressors such as nutrient pollution and overfishing may improve the resilience capacity of marine communities. Adopting climate-savvy fisheries management strategies that consider the spatial and temporal extent of hypoxia and are reflective of the changing nature of the marine environment, would aid in protecting oxygen-sensitive marine resources.

Abandoned Mine Drainage

Abandoned Mine Drainage (AMD) is water that has become contaminated as a result of historic coal mining. Some people call AMD "acid" mine drainage because it is commonly acidic, however it can be alkaline in nature so the more proper term is "abandoned". Prior to 1977, the laws governing coal-mining operations were less stringent concerning their environmental impacts. It was a common practice to simply abandon mining operations following the exhaustion of the coal reserve, and then declare bankruptcy. This allowed the mining operators to walk away from liabilities, including environmental devastation.

The nature of AMD contamination varies greatly from site to site, as its formation is dependent on a variety of factors. AMD often lowers water quality and impairs aquatic life, and is most often characterized by one or more of the four major components:

- Low pH (high acidity), i.e., acid mine drainage

- High metal concentrations (iron is the most common, but aluminum and manganese are also commonly found)

- Elevated sulfate levels

- Excessive suspended solids or siltation

Aluminum precipitate is seen in a stream.

Aluminum is toxic in the dissolved and precipitated form. B) Iron sediment is seen coating a small stream. Along with the iron precipitate is algae which can coat impaired streams. C) A kill zone is an area of land where no vegetation grows. This is due to polluted water running over the surface. D) A polluted stream enters the West Branch of the Susquehanna. Iron and aluminum can be seen.

Formation of AMD

AMD is created when water and oxygen comes in contact with certain minerals which contain acidic material. In the case of AMD this happens in the mining process, but these minerals can be exposed in other ways. The mineral responsible for the vast majority of AMD formation is pyrite, often called fool's gold. Pryite's chemical formula is FeS_2 and is also called iron sulfide or iron pyrite. While in actuality a series of chemical reactions occur to form contaminated water, the net result of these reactions can be summarized as follows:

$$Pyrite + water + oxygen = sulfuric\ acid + yellow\ boy$$

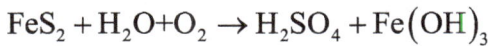

$$FeS_2 + H_2O + O_2 \rightarrow H_2SO_4 + Fe(OH)_3$$

Sulfuric acid (H_2SO_4) is a strong acid capable of having devastating environmental consequences for plants and animals. Yellow boy ($Fe(OH)_3$), also known as iron hydroxide, can form an orange or yellow sludge coating the bottoms of streams, effectively smothering aquatic life.

The acidity generated by this reaction can further dissolve other minerals and leach metals from the soils where the water flows. These metals most commonly are aluminum and manganese. These pollutants in sufficiently high concentrations can have a variety of negative effects.

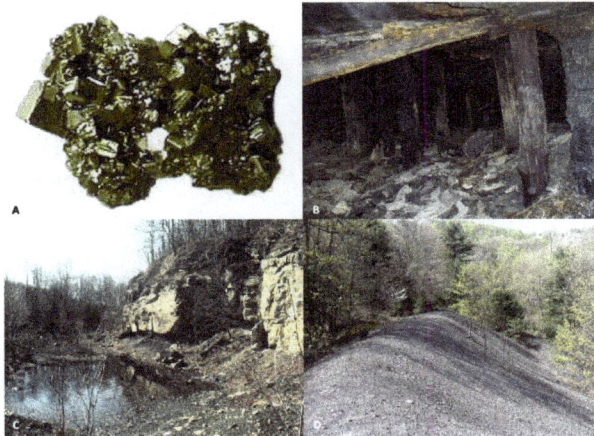

A) Pyrite is naturally found in the soils and near coal. As coal was mined Pyrite was exposed and has lead to AMD. B) Historic deep mines such as this one can fill with water and contribute large amounts of pollution. C) Surface mines were often left as open pits, and steep cliffs. These contribute pollution and are also a risk to people who may be injured in them. D). Unwanted rocks and low-grade coal were often times piled up and can still be found. These piles produce polluted water, as well as being eyesores.

Impact of AMD on Streams

AMD is responsible for degrading over 7,500 miles of streams throughout the Appalachian region, and has been identified by the Environmental Protection Agency as the single largest threat to the environment. This acid along with the metals in the water degrades water quality and affect life in the streams. Streams generally have very low pH (2-4) and high metal concentrations (iron aluminum and manganese), which leave streams discolored. Sulfates are also present which produce an un-

Impacts to Stream Life

- Low pH affects the ability of aquatic organism's cells to regulate how much water is contained in them. This leads to cell death especially in the gills.

- Dissolved and some precipitated metals, especially aluminum are toxic to life. When theses metals are combined with low pH they are more toxic.

- Metals coat stream bottoms leaving no habitat for macro invertebrates, and smothering the eggs of fish and macros.

- Precipitated metals coat and clog gills.

- Aquatic plants are lacking in these streams, limiting the food for macro invertebrates. As invertebrates die off food for fish is limited.

- Coldwater fish (Trout) are especially affected since small Coldwater stream generally do not have naturally occurring buffering capacity.

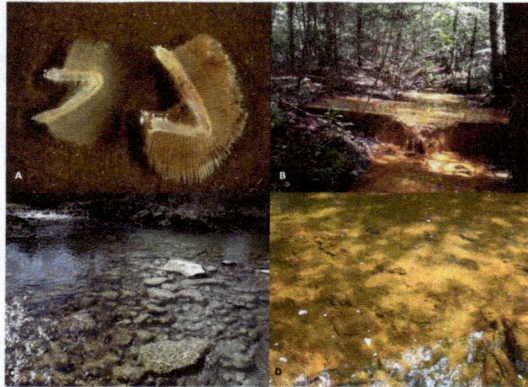

Impacts to the Economy

We are dependent on clean water, just as trout are. AMD impacts the water resources at our disposal and thus has massive impacts on human life around these streams. AMD also clogs pipes, and damages concrete structures in water. The following figures are a summary of the results of a 2009.

Recreational Spending

$22.3 million in sport fishing revenue was lost in one-year with-in the West Branch watershed. This does not include the losses in revenue from other eco-tourism and recreation.

Drinking Water Supplies

Actual assessments of how much money is spent treating AMD so it can be used in drinking water are difficult to assess. It is know that in the West Branch $11 million has already been spent to provide drinking water to contaminated wells, much more has been spent by public water companies to treat water.

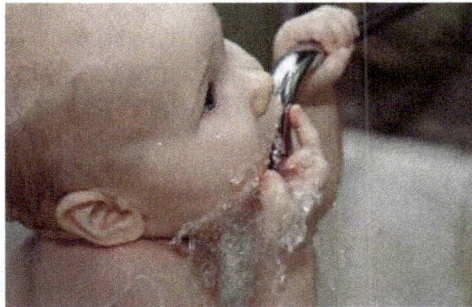

Property Values

In one small sub watershed of the Susquehanna $4 million dollars was lost in property values. Homes with-in 200 feet of an impaired stream lost approximately $2,500 per acre.

Job Creation

The restoration of streams will generate local jobs during the remediation process, and more jobs are generated with restoration of the area. Many areas where AMD impacts streams are economically depressed with little industry remaining.

Mitigation Strategies

Despite the vast scale and massive devastation that AMD has caused there are ways to restore streams so they can support life. Volunteer groups, state agencies, and other non-profit groups are working to install treatment systems, which remove the pollution and restore streams. Treatment involves increasing the pH of the water, which limits the ability of metals to be dissolved. Once the metals become solid they can be removed. We also can remove the source of the pollution through land reclamation.

Passive Treatment

A) Large treatment systems like this one contain multiple "cells" which treat water. If one of them is damaged or needs work, water can be diverted to the other cells allowing treatment to continue.

B) In this system pipes are placed at the bottom of the pond, with limestone over top. Mushroom compost is placed on top of to promote bacterial growth. Water must flow vertically through all the layers to the pipes on the bottom, so these types of systems are called vertical flow ponds.

C) Systems are often times quite large and utilize multiple types of treatment.

D) This type of passive treatment is calling a limestone cell and raises the pH of water.

Passive treatment involves using limestone, microbes, wetlands and other means to increase pH and remove metals. As the pH is increased by limestone (a very basic mineral) the metals become solid. These can then be settled in large ponds, and removed. Plants and microbes assist with this process. Each discharge is different so multiple different technologies are often times used. While these are called passive, they still require monitoring and maintenance every few months.

Fixing this Problem

Active Treatment

Active treatment involves adding basic chemicals to AMD to treat the pollution. These systems then allow the precipitated metals to settle and the clean water to enter the stream. These systems can be large water treatment plants, or just small "dozers" which add the chemicals. Active treatment involves continuous additions of chemicals, and requires monitoring and new chemicals to be added often. This generally means someone needs to be at the site at least once a week.

A) This is an example of a large water treatment facility. This facility cost $14.4 million, and requires a full time employ to maintain. B) A "lime dozer" is really just a silo filled with basic chemicals that regulates how much of this chemical is added at a time.

Land Reclamation

Active treatment involves adding basic chemicals to AMD to treat the pollution. These systems then allow the precipitated metals to settle and the clean water to enter the stream. These systems can be large water treatment plants, or just small "dozers" which add the chemicals. Active treatment involves continuous additions of chemicals, and requires monitoring and new chemicals to be added often. This generally means someone needs to be at the site at least once a week. Land Reclamation involves removing the source of pollution by re-grading and redirecting the water flow so that pyrite is not exposed. Basic material is also added to treat and acid that is formed. Sometimes coal that was not removed during the original mining process is also removed and sold to decrease costs, this is called remining. Land Reclamation permanently removes the source of pollution and requires little maintenance. It also improves the land habitat.

A) The Barns Watkins refuse pile and mine prior to reclamation. B) The same site after land reclamation. The water flowing here is the West Branch of the Susquehanna.

Aquatic Toxicology

Aquatic toxicology can be defined as the study of the effects of potentially toxic chemicals on aquatic organisms, with special emphasis on the harmful effects. Historically, this discipline has used toxicity tests to identify the harmful effects. Standard tests evaluate dose-response relationships (toxicity at different concentrations) and mechanisms of action in a variety of organisms that are representative of different ecosystem niches. These tests may evaluate the response of individuals or populations to varying concentrations of the chemical. The dose-response relationship is based on the following three assumptions:

1. The response (toxicity) is due to the chemical administered.

2. The magnitude of the response (toxicity) is related to the dose.

3. There exists both a quantifiable method for measuring and a precise means of expressing toxicity.

Types of Effects

Effects may be of such minor significance that the organism can function normally. However, under stressful conditions (i.e., pH change, low dissolved oxygen, high temperatures, changes in hardness, etc.), the same chemical exposure may become very lethal. The toxicity of some chemicals may also be enhanced or mitigated in the presence of other chemicals. In addition to killing the organisms, some pesticides can have negative but non-lethal effects on individual organisms and populations, such as reduced reproduction, reduced mobility to escape predation, or alterations in behavior.

Toxicity Measurement and Estimation

One common measurement used to describe toxicity of pesticides to organisms is the LC_{50}, or the statistically derived concentration in water that can be expected to cause death in 50% of the animals exposed.

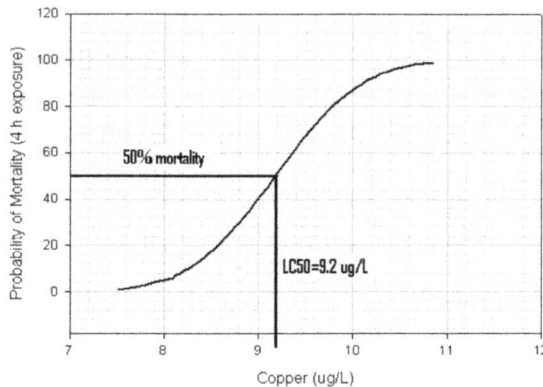

Figure: Sample toxicity curve showing water flea mortality. The sigmoidal line represents actual percent mortality. The LC_{50} = 9.2 µg/L in this example.

For estimation of non-lethal effects on processes such as growth and reproduction, the EC_{50}, or the statistically derived concentration in water that can be expected to cause a reduction of 50% in the process being measured, is used. Toxicity tests usually fall into one of two categories, acute or chronic. Acute tests are designed to evaluate the effects of pesticides on survival following exposures for a short period of their lifespan. Animals used in these tests are normally exposed for 24-, 48-, 72-, or 96 hours in order to estimate acute toxicity. In contrast, chronic toxicity tests evaluate effects over a significant portion (1/10th of lifetime or longer) of the organism's life span. These tests often evaluate sub lethal effects on reproduction, growth, and behavior, as well as mortality. Relative to acute effects, chronic effects may occur following exposures to lower concentrations of the pesticide. This chronic toxicity information is not always readily available because of the considerable expense associated with testing.

Standard Toxicity Testing Organisms

It is important to recognize that toxicity data will not always be available for all potential species in a given environment.

Given this limitation, the overall objective of test organism selection is to choose surrogates that are representative of the major ecosystem components. Aquatic algae and plants are representative of organisms that convert sunlight to carbon-based energy (the base of the food chain). Invertebrate species such as scuds and water fleas feed on algae and decaying plant materials and bacteria. These organisms are an important food source for larger invertebrates and fish. Fish species serve as an important source of food for a variety of larger fish, birds and mammals. Fathead minnows and sunfish often represent temperate warm-water fish, and trout represent cold-water fish species.

Risk Estimation

The risk of toxicity between a given herbicide and aquatic organism depends on the organism-specific, inherent toxicity of the compound and the concentration and duration of the exposure. The inherent toxicity is associated with the presence of a specific mode-of-action for causing toxic effects, and cannot be changed.

Herbicides labeled for application directly to aquatic systems or to ditch banks may be of special concern in aquatic systems because of the more direct exposure routes, compared to applications in a terrestrial situation. While the labels for these herbicides have been formulated to minimize potential toxicity in treated water bodies, it is important to recognize that there is a margin of safety associated with use of each at the labeled application rates; and that margin of safety can be significantly reduced with improper use.

References

- Water-pollution: eniscuola.net, Retrieved 26 April 2018

- The-importance-of-wastewater-treatment-environmental-sciences-essay: ukessays.com, Retrieved 24 June 2018

- Causes-and-effects-of-ocean-pollution: conserve-energy-future.com, Retrieved 18 July 2018

- Ocean-deoxygenation: oceanscientists.org, Retrieved 25 May 2018

- Aquatic-Toxicology-Notes-Predicting-the-Fate-and-Effects-of-Aquatic-and-Ditchbank-Herbicides: agrilife.org, Retrieved 21 June 2018

Permissions

All chapters in this book are published with permission under the Creative Commons Attribution Share Alike License or equivalent. Every chapter published in this book has been scrutinized by our experts. Their significance has been extensively debated. The topics covered herein carry significant information for a comprehensive understanding. They may even be implemented as practical applications or may be referred to as a beginning point for further studies.

We would like to thank the editorial team for lending their expertise to make the book truly unique. They have played a crucial role in the development of this book. Without their invaluable contributions this book wouldn't have been possible. They have made vital efforts to compile up to date information on the varied aspects of this subject to make this book a valuable addition to the collection of many professionals and students.

This book was conceptualized with the vision of imparting up-to-date and integrated information in this field. To ensure the same, a matchless editorial board was set up. Every individual on the board went through rigorous rounds of assessment to prove their worth. After which they invested a large part of their time researching and compiling the most relevant data for our readers.

The editorial board has been involved in producing this book since its inception. They have spent rigorous hours researching and exploring the diverse topics which have resulted in the successful publishing of this book. They have passed on their knowledge of decades through this book. To expedite this challenging task, the publisher supported the team at every step. A small team of assistant editors was also appointed to further simplify the editing procedure and attain best results for the readers.

Apart from the editorial board, the designing team has also invested a significant amount of their time in understanding the subject and creating the most relevant covers. They scrutinized every image to scout for the most suitable representation of the subject and create an appropriate cover for the book.

The publishing team has been an ardent support to the editorial, designing and production team. Their endless efforts to recruit the best for this project, has resulted in the accomplishment of this book. They are a veteran in the field of academics and their pool of knowledge is as vast as their experience in printing. Their expertise and guidance has proved useful at every step. Their uncompromising quality standards have made this book an exceptional effort. Their encouragement from time to time has been an inspiration for everyone.

The publisher and the editorial board hope that this book will prove to be a valuable piece of knowledge for students, practitioners and scholars across the globe.

Index

www.ingramcontent.com/pod-product-compliance
Lightning Source LLC
Chambersburg PA
CBHW061952190326
41458CB00009B/2854